HIDDEN WEALTH
The Survival Strategy of Foraging Farmers
in the Upper Arun Valley, Eastern Nepal

HIDDEN WEALTH
The Survival Strategy of Foraging Farmers
in the Upper Arun Valley, Eastern Nepal

By

Ephrosine Daniggelis

The Mountain Institute

Mandala Book Point

HIDDEN WEALTH
The Survival Strategy of Foraging Farmers
in the Upper Arun Valley, Eastern Nepal
Published by
Mandala Book Point
Kantipath, Kathmandu, Nepal
E-Mail: books@mos.com.np, mandala@ccsl.com.np
in collaboration with
The Mountain Institute
Asian Regional Office, Kathmandu, Nepal

Further information about The Mountain Institute is available from:
The Mountain Institute
E-Mail: summit@mountain.org

Visit The Mountain Institute website at http://www.mountain.org

This publication is financed with partial support from the
Royal Government of the Netherlands.

ISBN 99933-10-09-3

Photographs by Stephen Bezruchka and Ephrosine Daniggelis
Cover photograph: Rai Shepherd carrying *mangan ko sāg* by Ephrosine Daniggelis

Edited by Patricia Roberts

Layout by
Dongol Printers
Gophal Tole, Kathmandu, Nepal
E-Mail: dongolpr@mos.com.np

Preface

Mountain people have long practiced sustainable resource management. They have had to, because their place-based cultures and lack of integration in mainstream economics have required it. So too, the mountains themselves make integrated, intensive approaches the norm. The extensive development model of the lowlands simply is not practical in the variable topographies and climates of upland regions. Sustainability in mountains, then, is both art and applied science. What learning can the environment and development community glean from the wisdom of these traditional practices?

Ephrosine Daniggelis spent twenty months living and working among the Rai and Sherpa communities of the upper Apsuwa Valley, learning from them about their sustainable uses of the *jangal*, a mosaic of natural forests and agricultural fields in a little known part o Nepal. Her work is a landmark in furthering our understanding of the complex survival strategies of mountain people. Understanding how local people value and manage their own natural resources is the first step in designing new models of sustainable resources management ... something that. The Mountain Institute is committed to and has been doing for almost 30 years in the Himalaya, the Andes and the Appalachian ranges.

It is a pleasure for The Mountain Institute to co-publish this work, which adds importantly to learning about sustainability strategies in mountains. The research was supported by the Makalu Barun Conservation Project, an innovative project of HMG Nepal and The Mountain Institute to help local people manage and protect a globally significant natural treasure just east of the Mount Everest eco-system. The Mountain Institute has been working with the government of Nepal and especially with local people to help document these important indigenous knowledge systems. It is our hope that disseminating this knowledge will help conserve these and other biologically rich mountain areas, while creating opportunities for mountain people to earn sustainable livelihoods.

D. Jane Pratt, Ph.D.
President and Chief Executive Officer
The Mountain Institute

Preface

Mountain people have long practiced sustainable resource management. They have had to, because their place-based nature and flow of resources in mainstream economies have required it. So too, the mountains themselves make integrated, intensive approaches impractical. The extensive development model of the lowlands simply is not practical in the variable topography and climate of upland regions. Sustainability in mountains is, then, to both art and applied science. What farming can the environment and development community glean from the wisdom of these traditional practices?

Ephemeral though the past twenty months of research and writing seem, the Kid and Sherpa communities of the upper Mustang Valley, learning from them about their sustainable uses of the context, a measure of agriculture and agricultural fodder... as of the emotion's chemical materials of... nature... local people's culture and a range of their own natural resources, is the first step leading to sustainable resource management. It supplement that The Mountain Institute is committed to and has been doing for almost 30 years in the Himalaya, the Andes and the Appalachian range.

It is a pleasure for The Mountain Institute to bring this work, which adds immeasurably to learning about sustainability strategies in mountains. The research was supported by the Mustang Baraha Conservation Project, an innovative project of HMG/Nepal and The Mountain Institute to help local people manage and protect globally significant natural resources just east of the Mt. and Everest ecosystem. The Mountain Institute has been working with the government of Nepal and especially with local people to help them in these important indigenous knowledge systems. It is our hope that this continuing knowledge will help conserve these traditional biodiversity-rich mountain areas, while creating opportunities for mountain people to earn sustainable livelihoods.

D. Jane Pratt, Ph.D.

President and Chief Executive Officer
The Mountain Institute

Acknowledgements

I am indebted to the many persons and institutions who have supported me throughout my research project and graduate career. I regret that it is impossible to mention everyone by name.

First I would like to thank the members of my graduate committee. Dr. Alice Dewey, my chair, who incessantly extended support, inspiration, encouragement and kindness throughout the years and has always kept her office door open. I also thank my other committee members: Dr. Leslie Sponsel, who has challenged me through the years, been very supportive and enthusiastic; Dr. Bion Griffin for his critical advice and guidance; Dr. Jefferson Fox for his encouragement and expertise in Nepal's natural resource management systems; and Dr. Bluebell Standal who has provided valuable insights into the similarities between the Rais and her own people, the Khasi of Meghalaya.

Particularly helpful was Dr. Michael Dove, who critically reviewed Chapter 9 and has always inspired me with his thought-provoking questions and insights into local people's problems and solutions. Dr. Ramesh Khadka read Chapter 8 and provided helpful suggestions. Dr. Gregory Maskarinec was extremely helpful with the Nepali transliterations in the text and inspired me through the years. Dr. Carol Carpenter has always been generous with her suggestions and editing skills. Drs. Gerald Berreman, Robert Fisher, Martin Gaenszle and David Zurick were especially helpful during conversations and with their comments. And *mahalo* to D. Kyle Latinis, who spent many long hours scanning the farmer's natural resource participatory maps in Chapter 9. I also appreciate Vivian and Sonja's emergency computer help in the final stages.

My graduate studies in Hawaii have been generously supported by the East-West Center Program on the Environment, which provided financial and logistic support as a grantee student, research intern and during my write-up phase. At the Program on the Environment, I especially thank former Director Dr. Terry Rambo, Program Director June Kuramoto, and the support staff's *aloha*. The Department of Anthropology, University of Hawaii, has provided me academic guidance, logistical support, a tuition waver and travel fund award during my graduate studies. *Mahalo* to Mary Tugaoen and Elaine Nakahashi for always watching out for me. In my earlier years, the late Ethel Okamura was very helpful.

I would like to acknowledge the generous financial support provided by the National Science Foundation (Grant No. BNS-9020186) and Winrock International

Forestry/Fuelwood Research and Development Project (F/FRED) to conduct the research in Nepal.

I wish to thank His Majesty's Government of Nepal for granting me permission to conduct research in Nepal. I am grateful to my institutional sponsor, Dr. Dilli Ram Dahal and The Centre for Nepalese and Asian Studies (CNAS). Dr. Dahal always graciously invited me to his home to celebrate the Nepalese festivals with him and his family. Special thanks to Dr. Tirtha Bahadur Shrestha for his ethnobotanical expertise and support. *Tāshi deleg* to Tinley Dhondup who graciously helped with transliterations of the Tibetan words in the text. Thanks to Makalu-Barun National Park and Conservation Area project for approval of the research and logistic support. Especially I would like to acknowledge Dr. Gabriel Campbell, Bob Davis, Chandi Chapagain, Ang Rita Sherpa, Khagendra Sangam, and the late Tashi Lama.

Special thanks to Steve LeClerq and Penny Dawson for sharing stories of the East and their hospitality and friendship, and Pralad Yonzon for his friendship, support and advice. Thanks is also given to Kay Norton, Siobhan Kelly, Charla Britt, Stephen Bezruchka, John Raintree, and Tracey Parker. In Hawaii, many friends have helped me overcome numerous hurdles. *Mahalo* to Tricia Allen, Carolyn Cook, Rose Schilt, Jeff Burgett, D. Kyle Latinis, Marilyn Li, Sonja Brodt, and Nancy Cooper.

Dhanyabād to my research assistants, Ngima Temba Sherpa, Lila Ram Rai, Dawa Keema Sherpa and Sonam Gyamju Sherpa. Also, Ang Nurbu Sherpa, Bal Bahadur Rai, Gar Bahadur Rai and the late Dute *ko bāje* for his knowledge of the early settlement history and strong *raksī*. Chiring Dorji Sherpa, for his assistance with the medicinal plant trade and friendship. Pasang Phuti who led me on many paths to the *jaṅgal* to discover its wealth. In Kathmandu, Tsiring Sherpa always was willing to answer my numerous questions and feed me *momos* at the Makalu Restaurant. I am especially grateful to the Rai and Sherpa marginal hill farmers who graciously opened their homes and openly discussed their difficult lives. They taught me not only about the *jaṅgal*, but much more, and shared their hopes and dreams for a better future for their children.

The publication of this manuscript has been made possible by The Mountain Institute and the Global Environment Facility of the United Nations Development Programme. Thanks to Brian Peniston, TMI Representative, for choosing to include this work in the Natural Resource Management Series, and to Patricia Roberts, who edited the final manuscript and supervised printing and graphics work.

Last, I always remember my late beloved brother Spiro who inspired me to travel off the beaten footpath and explore the unknown. This dissertation is dedicated to Bijale whose tragic death occurred shortly before I left my research area. I knew it was time to leave. Bijale was my dear friend and teacher and in all of our hearts, "the greatest *dhāmī* of all".

Foreword

The contrasting perspectives on the complex mosaic of forests and fields called the *jaṅgal* reflect how local people use it differently from external agencies. The power-holders view the *jaṅgal* as "low status" and "peripheral," and their regulations affect the welfare of the local communities living on its fringes. Twenty months of research were conducted during 1991-1993 in the upper Apsuwa Valley, Eastern Nepal, among Rai and Sherpa marginalized hill farmers living on the *jaṅgal*'s fringes. Ethnographic, ethnobotanical and nutritional data were collected to explore the subsistence pattern of these farmers. The study showed that the *jaṅgal* is viewed by them as a polycrop in a broadly-conceived niche, containing resources imperceptible to outsiders. The *jaṅgal* is key to their survival, especially as available land has become reduced and more fragmented due to increasing population and indebtedness.

The *jaṅgal* provides an economic buffer by allowing swidden cultivation and the trade in *chiraito* (a medicinal plant) to alleviate the impact of food scarcity periods. Traditional swiddens increase, rather than degrade, biodiversity. Oral history links 'wild' tubers to Rai cultural identity. 'Wild' tubers are nutritionally rich, but are devalued by outsiders. Women depend on the gathering of such ingestibles to provide staples and micro-nutrients for their household food security. 'Wild' greens are often held in disdain by groups hoping to improve their social status. These groups often suffer from nutritional deficiencies as a result. The fact that the *jaṅgal* contains sacred forests and many deities helps ensure that the Rais and Sherpas will carefully protect its resources.

This study used an integrated approach to look at the non-farm environment as the ultimate polycrop system. This environment has high diversity and is a repository of many valuable nutritional, medicinal, economic (subsistence and cash), religious and cultural **major** resources. External agencies often see resources of the *jaṅgal* as minor and miss its wealth. The changing of boundaries by the government has threatened the farmers' access to common property resources. If the *jaṅgal*'s diversity is to be protected and the people's use of it assured, co-operation between the government and the local people is essential.

All Nepali words have been transliterated following the method of R.L. Turner (1980), and with the assistance of Gregory Maskarinec. Sherpa, Tibetan and Rai words will be identified by the letters "S", "T", and "R" respectively. Tibetan spellings for Sherpa words, when equivalents could be found, have been transliterated by Tinley Dhondup.

Contents

Contents

Tables & Figures

Tables

Figures

In memory of Bijale, my dear friend and teacher. In all of our hearts,
he was the *ṭhūlo dhāmī*.

CHAPTER 1

The Use of the *Jaṅgal*

This book documents how the Tibeto-Burman groups living on the fringes of the *jaṅgal* perceive, use and manage it. The word *"jaṅgal"* or "jungle" stirs up all sorts of thoughts of a place that is hostile, wild, impenetrable, and inhabited by wild animals, deadly snakes, blood-sucking leeches, etc. Many of our beliefs may have been influenced by the early images of explorers and geographers, who wrote vivid accounts of the "primitive people" hunting and living within the tropical forests. But surely for these "primitive" people living in or near to the *jaṅgal*, it has a different meaning and is not such an inhospitable place.

These local actors are foraging farmers living in the middle hills of Eastern Nepal, and belonging to two Tibeto-Burman groups, the Kulunge Rai and Khumbu Sherpas. For both groups, the term *"jaṅgal"* includes a broad sweep of the environment. *Jaṅgal* refers to common property areas, including all forest land, fallow swidden fields and most grass and pasture areas surrounding their villages up to the elevation of timberline, about 4,200 meters. Above the *jaṅgal* are the high alpine meadows which are not perpetually snow-covered, referred to by the villagers as the *lekh*. When I use the term *jaṅgal*, it often includes the *lekh* as well.

The *jaṅgal* has been chosen as the focus of this research because it contains different ecosystems which provide enormous biodiversity. Previous research has focused on the forest, without appreciating the inclusive complexity of the *jaṅgal* as defined by the Rai and Sherpa farmers and the *lekh* which they include for subsistence purposes. In previous decades in Nepal, researchers focused on timber and fuelwood consumption. Then the interest shifted to fodder (usually tree foliage), because of the large livestock population and its association with forest degradation. Until recently, a concurrent loss of other forest resources used by the rural communities was rarely alluded to. Beginning in the 1990s, there has emerged a "new" interest in "non-timber forest products," "alternative forest products," "non-wood forest products" and "minor forest products." These terms encompass fuelwood and fodder, as well as other resources, including food and medicinal plants.

At first, even I almost fell into the trap of focusing on a single natural resource, that is, 'wild'[1] foods gathered from the forest. This research topic initially came to mind when I read the findings from the first nation-wide health survey conducted in Nepal in 1965-66 by Worth and Shah (1969). The survey found that only eight of eighteen villages studied had Vitamin A intakes close to the estimated World Health Organization (WHO), and Food and Agriculture Organization (FAO) requirements. Seven of these villages were located in the rural hill and mountain region. I felt that these villages were closer to the forest and had more access to "wild greens," a source of Vitamin A.

When the Rai and Sherpa farmers opened my eyes to the variety of "hidden" wealth in the *jangal*, and shared their knowledge of it with me, I realized the *jangal* is more than the forest. Therefore I should not focus narrowly on a particular resource when I talk about the general survival of the foraging farmers. At first, the farmers were surprised that a foreigner was interested in the 'wild' foods they collected, because these are considered to be "famine foods," "poor people's food," and "polluted" by the Brāhmans and other Tibeto-Burman groups striving to raise their social status. Their importance has been devalued because they are viewed as easily accessible and "free."

I soon realized that the broadly-defined *jangal* and the *lekh* (high alpine pastures) have an extremely high biological diversity. Foraging farmers who experience periodic food shortages utilize the resources found in these areas in a great variety of ways. For example, the *jangal* is cut for swiddens, and after the cultivated crops are harvested, *chiraito* (*Swertia chirata*), a valuable medicinal plant, flourishes in the fallow fields for several years. *Chiraito* is a source of cash income and medicine for the remote communities.

Forestry projects should not only look at the trees in the "forests," but also at the *jangal*'s non-forest resources, which are of nutritional, economic, cultural, and religious importance for the foraging farmers and their families. Falconer and Arnold (1991) and Hoskins (1983) have asserted that the narrow focus on the value of trees, solely for fuel, has overlooked the multifold uses of the forest for food, medicine, fodder, building materials and income. Gunatilake, Senaratne and Abeygunawardena (1993) showed that the extraction of non-timber forest products is more pronounced among the impoverished, and declines as household income increases. Pimentel et al

---

1   Technically 'wild' contrasts with domesticated, the latter referring to species whose reproduction is controlled by people. Even though this implies that 'wild' plants are unaffected, or largely unaffected, by human activities, it is hard to support (Etkin, 1999, personal communication). Throughout the text I will use the term 'wild' to refer to those plants, fruits and other ingestibles found in the *jangal* which are 'wild', but may still have been intentionally or unintentionally affected by the foraging farmers' behavior (see Chapter 8, Swidden Agriculture). When the farmer's management of 'wild' ingestibles is known, then the term is referred to as 'semi-domesticated.'

(1997) provided an extensive ethnographic literature review of the important contribution and value of non-timber products to world food security, as a source of food, various products, income, and employment. However, in their assessment, these authors left out the cultural dimensions of these resources, neglecting the role of the religious realm as an important part of the overall food security situation and management of the forest.

Hecht et al (1988) warn that the economic value of non-timber forest products needs to encompass all aspects of use-value, subsistence and cash value alike, because for many impoverished members of the indigenous communities, and especially women, all economic aspects of non-timber forest products are important. Plotkin and Famolare's book, *Sustainable Harvest and Marketing of Rain Forest Products* (1992), focuses on the economic value of non-timber forest products (NTFPs) at the local and international level in the tropical rainforests of Latin America. Godoy and Lubowski (1992) suggest guidelines for the valuation of non-timber tropical forest products. They assert that the majority of studies of the economic valuation of non-timber forest products come from the Amazon, and studies from the tropical forests of Asia and Africa are needed. A study by Gunatilake, Senaratne and Abeygunawardena (1993) looks at the valuation of non-timber forest products extracted from the Knuckles National Park in Sri Lanka, and supplies the qualitative information needed to affirm the importance of the role of non-timber forest products in forestry sector development.

Although there are exceptions (see Fox 1995 and NWFPs 1995a), many approaches leave out the importance of the broader meaning of the *jangal* and the concomitant variety of resources. The *jangal* not only includes forest, but also pasture, scrub bush land, and fallow swidden fields which are found in temperate regions. Moreover, these resources have more than just economic uses. They are also important in the religious realm, and as a basis for ethnic identity which helps to positively regulate the management of the *jangal*. I argue that the *jangal* is central to the economic, cultural and religious existence of Rais and poorer Sherpas, and needs to be considered within a holistic perspective.

Only by making an ethnographic study of the *jangal* in the cultural context, could I begin to understand which resources are used and why they are overlooked by outsiders. I will try to predict the consequences of "green shopping," a term used by Dove (1994) to refer to what happens when outside interests view any *jangal* resource as valuable. An example of such a situation has occurred in the case of the extraction of *chiraito* (*Swertia chirata*) when a high market price for the plant caused export competition to put pressure on it. One local response to this problem was that the villagers changed land tenure patterns to protect *chiraito* on private land. The means by which changing market prices affects the local decision-making strategies of the farmer will be further discussed in Chapter 8.

In marginal areas, impoverished farmers face the challenging task of surviving amidst unpredictable and fluctuating environmental conditions and natural hazards. They have few choices to remedy their situation and with the increasing fragmentation of land are forced to rely even more on the *jangal* which has been their sustenance for centuries. I look at how the farmers' view of the *jangal* affects how they maintain an ongoing mutualistic relationship with it. When local knowledge is incorporated into forestry projects and local actors are involved in the implementation process, maldevelopment need not occur and the situation for the already-impoverished farmer has hopes of improving.

The Research Hypotheses

My original hypothesis was that *jangal* foods are nutritionally important. This research showed that the *jangal* in its wider definition, which includes the *lekh*, is important not only for food, but in many other ways, not all of which are appreciated by outside actors. It became clear during my research that the *jangal* is ritually important to the people. I began to appreciate the importance of their religious views, which enjoined respect for the *jangal* and elicited conservation attitudes. What emerged from my study was the realization that there are indigenous management strategies for reducing the degradation and destruction of the forest, and more broadly, the whole *jangal* and its resources.

The conceptualizations inspired by Western scientific perspectives are too narrow. In researching the use of the 'wild' environment, ecologists have focused on the forest and looked at single activities such as cutting timber, collecting fodder, or gathering of medicinal plants. These resources are assessed for their value to the local population. The following five hypotheses address the shortcomings of such studies.

Hypothesis I
To understand the true significance of the 'wild' environment to the local people, particularly the poorest among them, it is necessary to look at the 'wild' environment which lies outside of village land and cultivated fields. The local people use the term *jangal* for this broadly-conceived niche. For the purposes of this research, I also include what is called *lekh*, the high alpine pastures.

Hypothesis II
In order to evaluate the value of the polycrop which the *jangal* and *lekh* provide to the local population, you have to take the broadest possible view of the uses made of this polycrop, because the *jangal* and *lekh* provide timber, fodder, food, firewood, medicine and ritual items. To focus on one aspect of use is to miss how critical these resources are for the survival of the poorest families.

Hypothesis III

In analyzing the patterns of foraging, the researcher must take into account that any expedition to the *jaṅgal* or *lekh* is an opportunistic, multi-purpose venture – just as the environment is a polycrop. That is to say, a trip to gather firewood is also an opportunity to pick 'wild' greens and collect medicinal plants for household use and as a source of income. As a result, the studies focusing on such things as time spent in gathering firewood do not represent the efficiency of use of time and labor of the local householders.

Hypothesis IV

The *jaṅgal* (including the *lekh*) has become increasingly important to poor Rai and Sherpa farmers as their land has become reduced and more fragmented due to population pressure and indebtedness, and because government regulations have restricted or may restrict access to these resources.

Hypothesis V

The poor Rai and Sherpa farmers use the *jaṅgal* and *lekh* in a sustainable manner.

A Historical Perspective

Subsistence Use of the *Jaṅgal* and *Lekh*

Historically the Rais and other Tibeto-Burman groups have foraged for *jaṅgal* resources on communally-held lands as a major subsistence strategy. Restrictions placed on the use of the *jaṅgal* resources by the ruling class of "outsiders" is one of the factors behind their modern impoverishment. The ruling classes' cultural view of the *jaṅgal* – as only a forest providing valuable timber, musk, etc. – has been of substantial importance in the creation of policies that have degraded and removed it from the control of the Tibeto-Burman groups. I argue that the impoverished Rais and poorer Sherpas who view the *jaṅgal* as a positive part of their world, use and manage it in a sustainable manner. Contrasting cultural perspectives exist (Milton 1996), which affect how the *jaṅgal* is viewed, used and managed at different levels of authority and by groups of different socio-economic statuses.

This conservation attitude held by the foraging farmers is reinforced by the fact that increasing poverty forces them to depend on the *jaṅgal* even more for survival. Over the centuries, they have come to have a mutualistic relationship with the *jaṅgal*. Their extraction of the *jaṅgal* resources is done in such a way as to keep their impact on the *jaṅgal* low and controlled. For example, Rais (and poorer Sherpas) do not sell timber or use it to build their houses. Instead they use quickly renewable resources such as bamboo. Their practice of swidden agriculture encourages a species-rich ecosystem, as long as the fallow is of sufficient length, making them in Meilleur's (1994:270) terms, "biodiversity enhancers" rather than "biodiversity simplifiers." As political, economic and ecological circumstances have changed, the Rais have continued to develop, both on a communal and an individual basis, adaptive management techniques designed to reduce destruction of the *jaṅgal*. In some parts of Nepal, the *jaṅgal* cover has actually improved while simultaneously being exploited and cared for by the local communities (Fox 1993; Gilmour 1988; Ingels 1995). In addition, there are examples of new innovations in land tenure created by group consensus in my research area as a result of the medicinal plant *chiraito* (*Swertia chirata*) having become a valuable trade item in the past few years.

The Rai and Sherpa respect for the *jaṅgal* is evidenced by the rituals performed to honor the deities and spirits of the *jaṅgal*. It can also be seen in their myths and legends which tie their ancestral roots to the *jaṅgal*.

Environmental Concerns in the Nepal Himalaya

In recent decades, there have been alarming predictions regarding the ecological degradation of the Himalayan mountain ecosystem. A 1979 World Bank report predicted that because of population pressure and scarcity of land, all the accessible forest cover in the Hills and Tarai (plains) would disappear in the next fifteen and twenty-five years (respectively), "unless reforestation programs are initiated on a massive scale" (World Bank 1979:i). This "ecological crisis" led to the formation of the Himalayan Environmental Degradation (HED) theory which argued that recent (post-1950) population pressure in rural areas, insufficient landholdings, and poverty were linked with massive deforestation (Blaikie et al 1980; Eckholm 1976). Blaikie (1995:210) discusses how science has identified and framed these so-called 'environmental problems' which have adversely influenced policy-making and opinion formation.

Ives and Messerli (1989:9) have posited that the HED theory's linkages and assumptions were based on latter-day myths, falsely-based intuition, or unreliable data. For example, estimates of annual per capita fuelwood consumption differs by a factor of 67 (Donovan 1981). Several scientists have argued that the root cause of erosion in Nepal is natural: the geophysical features of the Himalaya (extreme altitude and steep slope) and the continuing process of uplift lead to a fragile unstable landscape, along with a naturally highly erosive climatic regime. It is also said that the subsistent Himalayan farmer has wrongly been made the "scapegoat of this ecological crisis" (Carson 1985; Ives and Messerli 1989).

While the geological cause may be the root of the problem, a number of authors point out that local conditions are influenced by overpopulation and complex socio-political factors which have an impact on the environment. Using a historical and political perspective, Bajracharya (1983) argues that particular policies reduced the farmer's holdings and so have become a cause of deforestation through the creation of rural poverty. The government policies have also undermined the local management systems, causing them to work less efficiently and increase natural resource degradation. An additional impetus to resource depletion in the Nepal Himalayas has been attributed to the unplanned growth of tourism (O. Gurung 1994).

Long-term approaches and multiple solutions which incorporate indigenous knowledge and directly involve local people are therefore required (Ives 1987; Ives and Messerli 1989). O. Gurung (1994) stresses the importance of conducting microstudies at the local level. These studies illustrate the multiple variables of socio-economic, environmental and historical dynamics in the Middle Hills. One such study in the Khumbu region used aerial photography to look at whether diachronical changes have occurred in the forest cover. A comparison of photos in the past thirty years (see Byers 1987) along with soil profiles and dendrochronology showed that the forest cover in the Khumbu region has not significantly changed during this time period (Byers 1986).

The Politics Behind the Cultural Interpretation of the *Jaṅgal*

The Himalayan Degradation Theory (HED) has been the focal point of controversy in Nepal regarding the problem of deforestation. Several issues surrounding this theory are raised: What are its causes? And 'who' or 'what' is the culprit? If deforestation has been occurring at the alarming rates predicted, why are there still forests remaining throughout Nepal? And why are they in even better condition in some areas than had been observed in previous decades? Fox (1993) and Gilmour (1988) have attributed this improvement to the active participation by villagers in the local management of natural resources. This is not entirely a recent phenomenon, but has been occurring for centuries in the form of traditional safeguards (Arnold and Campbell 1986; Cronin 1979; Furer-Haimendorf 1964; Molnar 1981b; Stevens 1993).

This book considers the underlying issues of forest degradation within a historical-political framework. The state's regulation of forest resources has exacerbated a social asymmetry, one of the factors leading to the widespread poverty present in contemporary Nepal. The cultural meaning of the term *"jaṅgal"* has had a bearing on how the forest has been managed and used by external influences throughout the history of Nepal, as well as other countries within the South Asian region (see Alcorn and Molnar 1996; Dove and Carpenter 1992; Jodha 1994).

Francis Zimmermann's 1987 book *The Jungle and the Aroma of Meats* explores ayurvedic formulas for food remedies. Zimmermann identifies the original Sanskritic meaning of *jāṅgala* as "dry lands" and "open" vegetation cover inhabited by Indo-Aryans; and the cognate term *jangal,* which in classical Hindi means 'wild' land – "uncultivated, non-civilized," an area inhabited by tribal groups.

Dove (1992) interprets Zimmermann's analysis differently by historically following the evolution of the term *jāngala* into *jangal.* Originally *jāngala* was used for grassland and was positively valued by the early pastoral Aryan. In later times, the term *jāngala* dropped out of use and *jangal* became the name of 'wild' forests in contrast to cultivated agricultural land. *Jangal* was used to disparage natural resources such as those occurring in fallow areas, and the term *"junglee"* was used in a derogatory fashion to refer to forest people.

Dove (1992) documents the polarity which existed between the pastoralists who used savannah and the villagers who benefited from the fertile irrigated land. The forest peasants (non-Aryan tribes) were pushed into inhospitable zones of the non-*jāngala*. The most important point is that originally the *jāngala* was ritually "pure" and the home of the Brāhman, whereas the non-*jāngala* area was "impure" and uncivilized, the refuge into which the Brāhman drove the barbarians (Dove 1992:234). In later times, the Hindu rulers viewed nature and the *jangal* (as the term was now used) as polluted, to be conquered, controlled and changed, and were culturally detached from it.

The Sanskritic view parallels some aspects of a common anthropological approach to the culture-nature dichotomy. The culture: nature theoretical approach formulated by Ortner (1974) was built on the Levi-Straussian structuralist notion that culture and nature are universal categories – and because culture is superior, it can dominate, control and transform nature. Terms used to refer to the *jangal,* the contemporary concept of the non-*jāngala* area, have included "inferior," "marginal," "forbidden," "the abode of wild animals" and "inhabited by tribal groups."

On the other hand, Diemberger (1992:CH3,35) described the Sherpa view of the deepest forest as a dangerous place.

> The forest is symbolically opposed to the inhabited and cultivated area, it is the uncontrolled and uncontrollable. It is the place for secret meetings, illegitimate love affairs, etc. and all kinds of spirits roam there. The cremation ground, which lies in the forest above the village, is a place favored by the dead spirits who couldn't find their way. The defilement of death clings to that place together with an aura of sacredness.

There are three distinct views of the *jaṅgal* and its relationship to cultivated areas.:

1) The Hindu view is that the *jaṅgal* is "impure," uncivilized" and 'wild.' It is to be conquered and controlled. This attitude is still current.
2) The anthropological view follows Levi-Strauss's theoretical division between nature and culture. In this approach, culture is superior and possesses the power to dominate and transform nature.
3) The Sherpa view (which is similar to that of many other Tibeto-Burman groups) sees the forest as uncontrollable and sacred.

The *Jaṅgal* in the Arun Valley

The remote Arun Valley, where I conducted my research, typifies Dove's (1992:238) description of the *jaṅgal*'s location. The *jaṅgal* is found not in the central plains but in hilly peripheral areas whose remoteness or some other characteristic (such as the slope of the land) makes them relatively less suited to agriculture. The historical and political context of the attitude toward the ecological relationship of culture versus nature in Nepal is the same situation as has been described for India (Zimmermann 1987) and Pakistan (Dove 1992). Because of government policies, the Himalayan Tibeto-Burman groups have been pushed into more marginal and infertile land where maize and barley are the major unirrigated grain crops. On the other hand, the Indo-Aryan rulers occupied the lush fertile valleys and irrigated hill-terraces where rice cultivation is possible. Rice itself is a "high status" food and a symbol of the powerful elite which I discuss in Chapter 6 of this book. Another example of the cultural disdain held towards the *jaṅgal* by the "power holders" is Sanskritization, whereby the removal of *jaṅgal* edible foods from the diet is one means by which an individual strives to move up the hierarchical structure where Brāhmans are at the apex. McDougal (1979:4), points out the Rais (a Tibeto-Burman ethnic group) have avoided "Sanskritization" and have remained essentially "tribal" with a positive view of the *jaṅgal*.

The format of this book is as follows:

* Chapter 1 looks at center-periphery relations, specifically Nepal's forestry policies in a historical perspective and how the ruling class has controlled and managed forest resources adversely affecting the Tibeto-Burman groups.
* Chapter 2 outlines the ecological anthropology approach used to look at the cultural, religious, social-economic and nutritional significance of the extraction, use, management and flow of *jaṅgal* resources from the *jaṅgal* gate to the household and market.

- Chapter 3 describes the research setting, the demographic structure and historical settlement patterns of the villages.
- Chapter 4 illustrates the foraging farmer's household food security situation by using a seasonal calendar to show when food scarcity occurs, and the adaptive and coping strategies employed by the farmer.
- Chapter 5 looks at the farmer's local knowledge of the *jaṅgal* resources, the species, their location and the social groups who extract them.
- Chapter 6 looks at how the cultural and religious importance of these *jaṅgal* resources are embedded within myths, legends and rituals. The social and cultural factors influencing how the *jaṅgal* resources are distributed at the individual, household and village level, are discussed.
- Chapter 7 focuses on women's role in the collection, processing and consumption of *jaṅgal* resources, along with the cultural and social factors which make these women nutritionally vulnerable within the household.
- Chapter 8 discusses the trade in *jaṅgal* resources and their relationship to swidden cultivation and the household economy. Conflict mediation case studies illustrate the methods of extraction of *jaṅgal* resources used by various actors and the types of conflicts which occur.
- Chapter 9 focuses on different levels of natural resource management systems, sacred forests, local knowledge systems and the villagers' pressing social concerns.
- Chapter 10 summarizes the main issues of the thesis and addresses the research hypotheses. Suggestions of how to look at these problems through the local people's "eyes" and work together with them are discussed.

Why the Rais Perceive Themselves to be *"Jaṅgal* Owners"

The Rais, a Tibeto-Burman speaking people, consider themselves to be descendants of the Kirāta who were first mentioned in the ancient Hindu epic, the *Mahābhārata*, ca 500-200 B.C. (Mookerji 1956).

Hodgson (1880 in McDougal 1979) felt assured that the Kirāti people were "forthcoming in their present abode from 2,000-2,500 years back and that their power was great, reaching possibly at one time to the delta of the Ganges." During their dynasty , the Kirāta kings overran the Kathmandu Valley and ruled Nepal from around 110 A.D. (Mookerji 1956:399). Linguistic evidence shows that the Kirantis who inhabit Eastern Nepal are clearly related to the hill tribes of Northeastern India (English 1985). Prior to the Gurkha conquest in 1768, the Rais and Limbus were the majority population in the Eastern Hills but they are now the minority (Caplan 1970). Zangbu and Klatzel (1995) note that in one story of early Khumbu settlement, the Kiranti Rai were claimed to be the earliest human settlers before the Sherpas arrived about 450 years ago. The Rais' symbiotic relationship with nature is shown by

Sharma et al (1991:20) remark, "no other cultural group has been described in its forest settings more truly and consistently than the Kiranti tribes in the ancient Sanskrit literature."

Nepal's Forest Policies

A historical framework of the forest legislation policies in Nepal shows they have adversely affected natural resource management. Prior to the formation of the political nation-state of Nepal in 1769, the country was fragmented into eighty autonomous kingdoms. A majority of the forest in the upper midlands was used by indigenous "tribal" groups (Rai, Limbu, Tamang, Lepcha, Sherpa, Sunwar, Danwar and Majhiya) of Eastern and Western Nepal. These groups maintained communal responsibility under customary forms of tenure.

This land tenure came to be known later as *kipat* for the Rais and Limbus of Eastern Nepal. An individual obtained rights, but not ownership, to land by virtue of membership in a particular clan and also by the geographical location of the communal land. A similar system, referred to as *riraij*, was also practiced by the Khasi of Meghalaya, formerly the hills of Assam (Standal, personal communication, 1997). The basic needs of each family for firewood, fodder and timber could be met while sustaining the resource base. Forests and pasture lands were considered community property that could only be used by non-community members through the payment of fees or other commodities (Caplan 1970; O. Gurung 1994; Regmi 1978).

In 1769, Prithvi Narayan Shah, the Hindu king of Gorkha, West Nepal, came to power and founded the nation-state of Nepal, which unified more than half of the eighty autonomous petty states. Exploitation and alienation of land was rampant during his reign because land was the state's ultimate resource. The local resource management strategies within Nepal were disrupted due to the penetration of dominant Indo-Aryan groups into the hill and mountain regions. The Shah rulers gradually confiscated the land held in customary forms of tenure from the "fugitive people" (the Limbus who fled to India after the Gorkha conquest of 1774) and converted it to *raikar*. *Raikar* is state ownership of land where the individual has the right to utilize and transfer the land as long as taxes are paid. The high caste Indo-Aryan groups obtained large landholdings and Tibeto-Burman groups were alienated (Allen 1976a; Griffin et al 1988; O. Gurung 1994). *Tālukdārs* (tax collector functionaries) controlled access to some of the forests and collected revenue for the use of its forest products in the hill districts and certain areas of Kathmandu Valley (Regmi 1978). The clearing of forests for agriculture was encouraged to enhance land revenue (Bajracharya 1983), confer wealth and prestige on the ruling group and by farmers seeking tax concessions (Griffin et al 1988).

Another loss of Tibeto-Burman tribal land began as early as the 15th Century, in the form of *jāgir*[2] and *birtā*[3] land grants which helped to create a feudalistic class who gave social and political support to the rulers. Forest land was also held under *guthi*-trust, that is, land assigned to religious and philanthropic institutions. These beneficiaries, like the state, sought to increase their revenues by converting forest land to agriculture. For a long time, the rent or tax paid by tenant peasants was one-half of the agricultural yield. The state, however, could not always use agricultural products paid as a tax by tenants (*jagerā*) and the equivalent in cash had to be paid. This policy led to the beginning of the widespread indebtedness and debt bondage found among so many rural people today. Also stressful was the *rakam* system utilized by King Srinivas Malla (a Newar, one of the Tibeto-Burman groups living in Kathmandu Valley) in 1672. Under this system, the peasants were obliged to provide porterage and other services to the nobility. These obligations constituted an "onerous" burden on the poor peasantry which was only partly compensated by minor tax concessions, although tenurial privileges were not secure (Regmi 1978).

During the Rana regime, which began in 1846 and stayed in power for more than one hundred years, the Rana rulers continued the previous policies of extraction of forest resources viewed as profitable commodities. For example, *dāurā rakam*[4] workers were obliged to supply 360 *dhārni* (about 1,800 pounds) of firewood per annum in 1855 (Regmi 1978:515).

> An important category of goods required to be transported by *rakam* workers consisted of building timber used primarily in the construction of palaces for members of the Rana family and their relatives and favorites. This construction activity created such a large demand for timber that a special office was established to organize *rakam* services for this work (Regmi 1978:512).

Another *rakam* obligation, to supply *jaṅgal* grass and fodder, was referred to as *ghanse rakam* (Regmi 1978: 856), and further depleted the forests. All *rakam* obligations were abolished in 1963.

The forests deteriorated further with the transformation to a market economy, where an immeasurable quantity of hard timber was sold to British India for the construction and expansion of the railway. The process of the states eroding the traditional usufruct rights of the poor continued (O. Gurung 1994). The introduction of maize and potatoes (which grow under marginal conditions where few crops thrive)

---

2   *Jāgir* were temporary land assignments given to government employees "in lieu" of salary, were tax-free and remained valid only while the individual was serving the government (Regmi 1978:v).

3   *Birtā* land grants, tax exempt and private, were made to soldiers, servants of the state and the nobility, mainly to high caste groups, Brāhmans and Chhetris (Regmi 1978).

4   *Dāurā Rakam* was a *rakam* obligation to supply firewood (Regmi 1978:856).

in the early 18th Century allowed for clearing of more marginal (steeper and higher) unirrigated land.[5] By the end of the century, especially in Central and Eastern Nepal, the process of deforestation was already well underway (Griffin et al 1988).

Once the Rana regime was overthrown in 1951, further regulation of forest resources was imposed by the state in order to curb deforestation. In 1957, the government nationalized all communal forests and "waste land" with the Private Forests Nationalization Act, and the control of the forests became vested in regionally-based officials. The indigenous community resource management system broke down, restrictions on resource extraction were not enforced and the forests were no longer protected. There were no records of land ownership and villagers were more apt to destroy the forest since, once the land was cultivated, it could be claimed as private property (Bajracharya 1983 and Wallace 1987). By 1968, all *kipat* lands were converted into *raikar* and local people who were not able to register their land were forced to encroach on forested areas (O. Gurung 1994). These regulations were not enforced in the Limbu area, nor in my research area where the Kulunge Rai live.

Because of the negative consequences of the Private Forests Nationalization Act, in 1977 the Government of Nepal revised its forest policy and returned formal control and management of forests to the village *panchāyat*.[6] New categories of forests were established: *Panchāyat* Forests, *Panchāyat* Protected Forests (existing community-used government forests), Religious Forests and Leasehold Forests (Wallace 1987:5). "*Panchāyat* Forests were defined as degraded forest areas entrusted to a village *panchāyat* for reforestation in the interest of the village community. *Panchāyat* Protected Forests are existing forests entrusted to local *panchāyat*s for protection and proper management. Religious Forests are forests located at places of religious importance entrusted to religious institutions for protection and management. Leasehold Forests are the furthest step toward private ownership as they allow lessees to reap the benefits of afforesting and managing degraded land" (Wallace 1987:5). The emerging paradigm assumed the people would fully participate in rural development activities.

Researchers have debated on whether or not the externally-initiated sponsored systems set up to manage the forests at the local level have been effective. These externally-sponsored forest systems have had mixed results. Ingels (1995) remarked that community forestry has shown to be an effective model for encouraging the conservation of forests. Fox (1995) found an improvement in the forest after the

---

5   Crosby (1972) asserts that because the yield per unit of land of maize and potatoes is double that of wheat it had a tremendous environmental impact in the temperate world, including Europe and Russia.

6   A *panchāyat* was a community-level administrative and political unit in Nepal which consisted of about 4,000-6,000 people (Gilmour 1988:346).

introduction of a new government tenure regime and assistance by a non-governmental organization. Gilmour (1989) notes that the success or failure of community forest programs depends on the continuum of varying resource availability which determines the villager's need for natural resource management.

Messerschmidt (1995:242) asserts that the assistance of "outside help" in addressing forest development and resource management is needed due to population pressure, economic opportunities and other factors. However, success depends on "efforts that reflect continuity with and are guided by pre-existing forms of management, indigenous knowledge, local co-operation and popular participation."

On the other hand, Fisher (1989) remarked that the government's implementation of forestry user groups was not very effective for several reasons. He said that there has been a poor level of understanding of the organizational basis for collective action (inappropriate groups of forestry users) and the emphasis has now been towards viewing forest user groups as managers. Another problem has been that the notion of the *Tragedy of the Commons* has remained the ideology of the major funding agencies. Seeland (1997:10) took a harder stance when he remarked that "the constitution of a forest-user group is a demonstration of power by the Forest Department at the village level." He further argued that "the making of an 'artificial' community or 'synthetic' social unit thus becomes a development objective that may contradict social reality, in which social strata are already-existing dimensions of the multicultural fabric of Nepali rural life in the middle hills" (Seeland 1997:8). This is supported by King et al (1990) who discussed two case studies showing that the implementation of community forest management can work if it involves forest users (including women and low castes) in the decision-making process. King et al (1990), Pandey (1993), and Seeland (1997) assert that the composition of a user group should not be dominated by a particular ethnic group or caste. Pandey (1993) lists the factors affecting the degree of successful management of forest resources in community lands. According to Dove and Rao (1990), the creation of new institutions is more likely to fail than is the use of extant traditional institutions.

The Nepalese government's community forestry program is an attempt to set up a system of management of forests as common property (Joshi 1989). Indigenous systems for managing forests as common property were already being used by villagers throughout Nepal. The land which is treated as common property is actually government land (Fisher 1989). Caplan (1970) claims that in Limbuan, the *kipat* system has not been abolished, although the Limbu land has been reduced.

The current debate pertains to the parts of the Himalayan resource base which are communally exploited. Hardin's (1968) study of common resource management, *The Tragedy of the Commons* asserts that the institution of common property leads to the overexploitation of resources because individual benefits will exceed the costs. Dove and Rao (1990) point out that Hardin's analysis is in error because it assumes that the absence of sanctions is the norm, while on the contrary, sanctions are an important feature of traditional societies. Hardin's argument left out the role of institutional arrangements and cultural factors in governing access to and use of the resources (see Feeny et al 1990: McCay and Acheson 1987). It is "soundly refuted" in the hills of Nepal where villagers effectively manage resources as common property regimes (Gilmour 1989:13); also see Chhetri and Pandey (1992); Messerschmidt (1986). I will cite a few examples pertinent to the Nepalese context, which demonstrate that traditional institutions help maintain a balance of communal resources.

The *Kipat* System of Eastern Nepal

The *kipat* system, practiced in Eastern Nepal by Rais and Limbus, is a form of both traditional and indigenous forest management systems (Fisher 1989). The rights to maintain *kipat*, a customary and communal system of land tenure of the *jaṅgal*, was granted in the late 1700s by the Shah regime to the Rais and Limbus. In local terms, *kipat* means the system by which all "things" – that is, streams, rivers, grazing areas, steep cliffs and trees – are shared by all the villagers. Most *jaṅgal* resources ('wild' greens, dry firewood, bamboo and medicinal plants) are treated as common property by the farmers in the villages located in the Arun Valley, Eastern Nepal, where I conducted my research. The concept of *kipat* is still meaningful to farmers there because they view the forests as 'community forests' although they legally became government forests in 1957. The system of *kipat* was still practiced by the farmers in this area. However, they felt their rights to use of the communal land would change once the Cadastral Survey re-registered it.

Kipat land in the area of Majh Kirat (Middle Kirat), the area where my research was located, between the Arun River and the Dudh Kosi, was said to have disappeared "long ago" (see Figure 1). But the Sankhuwasabha Subdivision is an exception: although the area is situated west of the Arun River, it forms part of the Pallo-Kirat (Far Kirat, mainly inhabited by Limbus), between the Arun River and the eastern border of Nepal.[7] Pallo-Kirat is the only area where *kipat* lands had not been converted to *raikar* tenure (state ownership of land) and the *jimmāwāl* system[8] has remained strong (Regmi 1978:539). Even though the *kipat* system was abolished legally, in certain Rai and Limbu areas, it is actually still in force.

---

7   Krishna Prasad Bhandari, "Pallokirat Ko Jagga (Land in Pallo-Kirat), *Sumyukta Prayas*, Bhandra 8, 2016 (August 24, 1959). In Regmi (1978:539).

8   A village land collector or headman. Regmi (1978) defines *jimmāwāl* as a collector for *khet* (irrigated land). However in the research area, the farmers used the term to refer to tax collectors on their unirrigated land (*bāri* or *pākho*).

Fisher (1989) discusses the importance of differentiating between the terms 'indigenous' and 'traditional,' when describing the types of forest management systems in Nepal. "The term 'indigenous' refers to systems that are generated by internal initiative within a local community itself," while the term 'traditional' implies antiquity (Fisher 1989:3). The author uses the example that an institution can be indigenous ("native-born") without being long-established and also a system can be traditional, but not necessarily indigenous.

In remote forested areas and where forestry personnel have not been posted, traditional forest management systems have survived (Molnar 1981b). An indigenous and traditional system of resource management of the remote forest at Dahbaley was described in *The Arun, A Natural History of the World's Deepest Valley.*

> By mutual agreement everyone in the village shared the right to use the forest as they needed, but no one was allowed to clear the land. To promote a sustained yield, the headmen of the village assign rights to gather firewood in certain areas of each woodlot, and households jealously guard their territories; many territories represent traditional claims that date back several generations. Trivial uses of wood are discouraged, and when a household needs a particularly large tree for a construction project, they must pay a sizeable sum to the village headmen. The fundamental concept of a renewable resource is also recognized, and the headmen will sometimes declare a moratorium on cutting if a certain plot shows signs of really excessive use that will soon lead to complete exhaustion (Cronin 1979:75-76).

Figure 1: Eastern Nepal showing the Kirat area.
Source: Mc Dougal (1979:5)

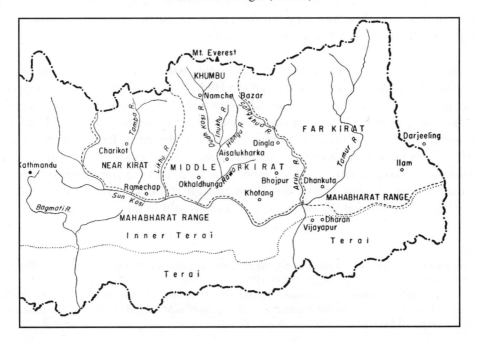

Stevens (1997:73) points out that oral traditions "testify" that sacred forests existed in the Khumbu area since the early years of Sherpa settlement dating back more than 450 years. The *naua*, village-based forest guards employed by the Sherpas of Khumbu, existed for centuries. The *naua* managed common forest resources by laying down extraction rates for fuelwood and timber (construction) and levied fines on villages that did not comply (Furer-Haimendorf 1964, Stevens 1993). See also discussion of sacred attributes of forests in the Upper Kali Gandaki (in Messerschmidt 1986).

A community forestry paradigm has emerged which points to the need for a subsistence farmer to be part of the forestry management planning process (Moench and Bandyophadhyay 1986; Thompson et al 1986). Malla (1992) stresses the fact that the subsistent farmer has been dynamic and adaptive in responding to socio-economic and environmental changes. This characteristic is now being recognized. Gilmour (1988) used the example that hill peasant farmers in two districts of Central Nepal are responsible for an increase in tree cover on their farmland, at least during the past twenty years. The author asserts that because virtually all forests in the hills are use by poor farmers "extra-legally," improved management of all forests is needed and not only portions of degraded government land.

Only by addressing the issue of poverty among subsistence foraging farmers can the government's efforts to reduce the pressure on natural resources be achieved (Gilmour and Fisher 1991; Ives et al 1987:333). The Government of Nepal's Eighth National Development Plan (1992-1997) focused on agricultural development with the goals of poverty alleviation, sustainable growth and resource conservation (Karan and Ishii 1994:77). One can hope this had a positive influence.

The Complexity of a Universal Definition for "Minor Forest Products"

The terms, 'non-wood forest products,' 'non-timber forest products' and 'non-timber forest resources' are often used interchangeably to refer to "minor forest products." However, a closer scrutiny of these terms shows that each one has its own nuances and characteristic classification categories. For example, the term 'non-wood' excludes materials extracted by rural people for fuelwood, building poles and small wood needed for handicraft and tools. In addition, insects, birds, fish and mammals may be left out of this definition. The term 'non-timber forest products' encompasses all biological materials, other than timber, which are extracted from natural forests for human use. Foods, medicines, spices, essential oils, resins, gums, latexes, tannins, dyes, ornamental plants, wildlife (products and live animals), fuelwood and other raw materials, notably rattan, bamboo, smallwood and fibers are included in this definition (deBeer and McDermott 1989:17).

Another term is "forest." Does forest mean primary or secondary forest? Must trees be included? And what about grassland areas? The impact of tribal agricultural practices on biodiversity is to create secondary forests which produce forest resources that never grow in other zones, an adaptive effect of succession (Cadelina 1985). The study of Malhotra et al (1993) found that even highly perturbed and degraded forests can still be rich in biodiversity. Many food sources and medicinal plants are gathered from secondary forests, particularly swidden fallows and along forest edges. DeBeer and McDermott (1989:106) warrant that "The differences in the particular array and abundance of useful plant and animal species in secondary forest as opposed to the original, undisturbed or 'primary' forest is a much under-researched topic."

DeBeer and McDermott (1989) note that human selection has modified species over centuries and it is often difficult (or impossible) to differentiate domesticated from 'wild' varieties. Some forestry sources use the term 'semi-wild' without defining it, while others refer to 'wild' as species or products of natural regeneration which are transplanted or captured from natural forests. Thus "the distinction between 'cultivated' and 'wild' food plants is not an absolute nor even a clear one" (deBeer and McDermott 1989:27).

These examples show that there is a great deal of ambiguity in the forestry literature regarding non-timber forest products. The term 'minor forest products' is often not defined. The products it includes are not listed nor is the ecological context described because it is assumed that the reader already understands the term. Whichever term is chosen, it needs to be adequately defined within the appropriate context in which it is used.

We keep fumbling around to find a proper definition for non-timber forest products until the meaning becomes too focused and leaves out the hidden importance for the farmer. I chose to study the cultural meaning of the term "*jaṅgal*" as seen through the eyes of the rural hill farmers in the Arun Valley of Eastern Nepal but to also substantiate its biological value using a scientific analysis. The Rais and poorer Sherpas use the *jaṅgal* in its entirety. It is viewed not only as a physical source of resources, but its ideological component is incorporated within the holistic framework of the environment.

Timber is used, but it is controlled for subsistence purposes. 'Wild' plants, fruits, mushrooms, medicinal plants, fish, birds and insects in these areas are classified as *jaṅgal* resources. The members of neither the Rai nor Sherpa communities normally hunt meat as a subsistence strategy because gathering non-game *jaṅgal* resources is of primary importance. However, bird species and the products which are used by members of both communities are included in the definition. In other words, *jaṅgal* plants, whilst primarily associated with a forest habitat, also include 'wild' plants of forest origin now growing in open fields or on pasture land without forest cover. My holistic orientation in this research is on *jaṅgal* plants not only for their nutritional, economic and medicinal importance for the Rai and poorer Sherpa foraging farmer, but also for the cultural and religious meaning they hold for Rai self-identity.

The Historical and Contemporary Use of Forest Resources in Asia

The value of non-timber forest products has varied with historical periods, geographical location, and power structure. When a commodity becomes economically valuable, the elite often exploit the local people, using their labor to collect it.

For at least two millennia, forest resources have been used for subsistence, bartering and cash trading – by hunter-gatherers, forest farmers (a vast majority of shifting cultivators), and pastoral communities. The value of 'minor forest products' outweighed that of timber in the historical past (deBeer and McDermott 1989). The evolution of forest product trading and its importance as a source of income for the indigenous populations has been documented in the Malayan rainforest (Dunn 1975), Eastern Kalimantan, Indonesia, (Peluso 1983), and Western Ghats, India (Morris 1982). For example, in Eastern Kalimantan prior to 1936, minor forest products (bird nests, rattan, resins, camphor and tanning barks) produced higher revenues than timber. Collectors were primarily women, young children and the elderly, for whom wood-chopping was a strenuous and time-consuming task (Peluso 1983). Dunn (1975) conducted an ethnoecology survey among the Malaysian Orang Asli (indigenous hunters and gatherers) which described the great range of ecological knowledge used to exploit their forest resource assemblage.

"Early" Tamil poets have documented that from "earliest times," the forests of the Western Ghats were inhabited by tribal communities who had important trading links with nearby agriculturists. In addition to this, external markets were established with the Romans by these early Tamil kingdoms. "Important forest products"—live animals, sandalwood, ginger, pepper, cardamom, and myrobalam – were traded. When forest products became valuable, the indigenous communities were exploited by those in power, that is, dominant castes and government agents, and then forced to collect wax, honey, and firewood for temples without remuneration (Morris 1982). English (1985) asserts that hereditary Hindu overlords in the lower midlands in central Nepal forced the villagers to provide unpaid labor services (*rakam*), to capture wildlife, gather wax, honey and medicinal plants, and to cut timber in the Tarai jungle. These goods were exported to Indian market centers by the Rajas.

Even today, many indigenous populations worldwide depend on minor forest products for subsistence, bartering and cash trade (Fernandes and Menon 1987; Morris 1982). Indeed their dependence is often so great that the word "minor" ought to be "major" instead. The Hill Pandaram in Kerala State, South India, collect minor forest produce and barter them with local villagers for staples as a significant element of their subsistence economy. These forest resources ensure a varied diet which consists of a 60:40 ratio of 'wild' to cultivated foods. Animal protein, in the form of nestlings, eggs and small game, provides ten to twenty percent of the Hill Pandaram's total food intake. Hunting is the dominant subsistence mode only in the high mountains and may provide at least ten percent of the diet; while in other areas, gathering is of primary importance (Morris 1982).

The Hill Pandaram live in a rich forested environment, even though many earlier writers have pictured them as being "on the verge of starvation" (Morris 1982:97). Iyer (1968:119 in Morris 1982) commented that the Hill Pandaram "seemed to enjoy better health in the natural forest than that of many communities living a 'permanent life'." Morris (1982:108) says this is the reason why they don't want "to sever the links that hold them to the forest."

According to Raintree and Francisco's (1994) example from the Philippines, 46 percent of the total multi-purpose tree products production (including many non-wood forest products) were used for family consumption, ten percent were given to neighbors and relatives, while only 44 percent were sold. This exemplifies the general finding that among many indigenous communities the importance of non-timber forest resources for subsistence purposes outweighs market opportunities. Approximately sixty percent of the production of non-wood forest products in India is consumed by about fifty million tribal people; and forty percent of their household earnings are from non-timber forest products (Lintu 1995b).

Fuelwood

Nepal is dependent on forest resources because nearly 93 percent of Nepal's labor force is rural-based and practices subsistence agriculture (Sharma 1989; Amatya and Newman 1993). Trees (fuelwood) comprise greater than ninety percent of the total biomass fuel requirement for the Nepalese population (Ives and Messeri 1989). The locality of collection is determined by economic status, as Fox (1984) noted, poor families in the low elevation village relied primarily on common lands for fuelwood and fodder needs, while wealthy farmers had a sufficient number of trees on private land. Fuelwood consumption rates are influenced by family size, caste and season. In addition to being used for cooking meals, fuelwood is also needed: in the preparation of *kũro* (mash used in animal feed); for religious celebrations; space heating, especially during the winter months; the distillation of alcohol (Fox 1984) and roasting grains. Alcohol production, a source of income for some Tibeto-Burman women, depends on the availability of fuelwood for distilling (Acharya and Bennett 1981). Other biomass fuels consist of corn cobs and stalk residues burned after the maize harvest in November and December. Firewood collection involving the felling of trees (freshly cut wood) is primarily, although not exclusively, the responsibility of men. Baragaonle males often go on overnight trips to distant forest areas to gather logs (Schuler 1981). Women usually collect dead branches and twigs and crop residues (Bajracharya 1980).

Ingestible Plants

Etkin and Ross (1991:231) use the term 'ingestible plants' to encompass both 'food' and 'medicinal' plants in order to study the biocultural adaptations made by the local people to their environment. Both medicinal plants and foods are considered to have healthful properties preventing disease by strengthening the body's defense mechanism. For example, in the Philippines, *dita* (*Alstonia scholaris*), a decoction of young leaves from a tree belonging to the *Apocynaceae* family, is used to cure beriberi (Neri 1994), caused by a lack of thiamine in the diet.

In Nepal, *jaṅgal* ingestibles include: 'wild' leafy greens, fruits, spices, condiments, grasses, bamboo shoots, seeds, nuts, roots and tubers, 'wild' birds, 'wild' animals, fish, insects and honey. Salt can be made by burning certain forest leaves and roots together (Alcorn 1996). *Jaṅgal* ingestibles are excellent nutrient sources, supplying vitamins, minerals, protein and trace elements. For example, 'wild' leaves may provide Vitamin A, ascorbic acid, folic acid, iron, calcium or niacin, and plant protein, depending on the species. 'Wild' seeds and nuts are sources of energy, essential oils, minerals and protein (deBeer and McDermott 1989; Kuhnlein and Turner 1991). Tubers, roots, rhizomes and corms provide starch, and the skins provide some minerals. 'Wild' fruits have higher concentrations of Vitamin C than exotic and domestic fruits (Wehmeyer 1966) and are good sources of carbohydrates, calcium, magnesium and potassium. Vitamin C enhances the absorption of iron and is important in countering the anemic conditions among women of child-bearing age, a major health problem in Nepal (Adhikari and Krantz 1989).

Scoones and Pretty (1992 in Alcorn 1996) found that in worldwide studies of forest dwellers it is common for forest foods to contribute up to eighty percent of their daily nutritional intake, with a higher percentage of these foods consumed by women, children, and the poorer population. Forest foods are of nutritional significance in the diet of reproductively active women, pre-adolescent girls and children, who tend to be at greater nutritional risk than adult males. Panter-Brick (1989) observed that pregnant and lactating Tamang women do not lower work output during the monsoon period, when the need for labor is increasingly high. Piller (1986) reports that Bajracharya's 1978 survey indicated that women of reproductive age work longer hours and eat less than adult males.

Piller's (1986) research found that while energy intake was sufficient in the diet for adult males and non-pregnant, non-lactating females, it was not adequate for lactating females. Hence Vitamin A was a limiting factor in the diet for all groups. Piller cautions researchers not to be "proteincentric" because the real cause of malnutrition may be overlooked. As mentioned previously, the first nationwide health survey conducted in Nepal (1965-1966) found that seven of the eight villages (out of eighteen villages studied) which had Vitamin A intakes close to the estimated requirements were located in the rural hill and mountain region (Worth and Shah 1969). According to a 1987 UNICEF (1987:83) report, the prevalence of xerophthalmia (a symptom of Vitamin A deficiency) was three times higher in the Tarai than in the Western Hills. It stated that no clear explanation for higher prevalence rates in the Tarai is currently available. As a consequence of this, there are two major ongoing Vitamin A projects being conducted in the Tarai.

However, at the National Vitamin A Workshop (1992:3-4), it was reported that high prevalence rates of xerophthalmia are also occurring in Jumla, a mountainous district in the Far West, Nepal. In Jumla, the problem of Vitamin A deficiency has warranted public health interventions. Further investigation in other hill and mountain villages is needed to determine if a similar problem exists. The greater abundance and consumption of *jaṅgal* greens may be an important factor for the lower prevalence of xerophthalmia or signs of Vitamin A deficiency in the hills and mountains.

In a 1977 study in the Eastern Hills, conjunctival xerosis (Vitamin A deficiency) was most often seen at the start of the monsoon, coinciding with the peak prevalence of diarrheal diseases (Nabarro 1984). Xerophthalmia occurring among pre-school children is often associated with malnutrition, diarrhea and measles (UNICEF 1987). Thus *jaṅgal* ingestibles can provide crucial amounts of essential nutrients, including trace elements which may not be present in cultivated or purchased foods.

Besides their nutritional contribution, *jaṅgal* ingestibles serve as buffers to seasonal and emergency food shortages for the marginalized hill farmer. They help reduce the risk of disruption of household food security during periods of stress (McElroy and Townsend 1985; Metz 1989; Piller 1986). Community household food security is defined as the "access by all households at all times to enough food for an active and healthy life" (Velarde 1990:13). *Jaṅgal* ingestibles provide staples, often referred to as "famine" foods; supply essential nutrients in the form of side dishes and snacks; and treat illnesses (traditional medicine) in remote and marginal rural communities (Falconer 1990).

The monsoon season, when food supplies are low, is also the period for peak agricultural activities. An adequate food intake is of vital importance to meet increased energy demands. When staple foods are in short supply during the monsoon period, the Kham Magar, northwest Nepal, traditionally gather "forest" food resources, mushrooms and 'wild' greens, which are made into a curry to accompany boiled grains. During the winter months, "nettles" are the main vegetable eaten by them (Molnar 1981a). *Jaňgal* foods can even be substituted for major staples. Mahua flowers are a seasonal grain substitute in India (Lintu 1995a).

Researchers have noted that *jaňgal* ingestibles are frequently consumed between meals by children (McElroy and Townsend 1985). This is important given their specific nutritional contributions for growth and development. Piller (1986:76) says in his research in eastern Nepal that one of the children "displayed a wide knowledge of, and a great fondness for, many of the 'wild' seeds, nuts and fruits available along the trail." Local people, however, may not consider *jaňgal* fruits and 'wild' greens as "foods," especially when they snack on them away from the household hearth.

Traditional medicine is still important in isolated and rural areas and *jaňgal* resources are an essential ingredient. An estimated eighty percent of the population of Nepal depends on traditional medicine as their sole or primary form of health care (Rajbhandary and Bajracharya 1994). In Nepal, approximately seven hundred medicinal and aromatic plants have been identified (Rawal 1994). Western 'modern' medicine is very costly, primary health care centers may not exist in the area, or when they are available, villagers do not have time to seek health care and traditional medicine is preferred. The rural communities often use their own home remedies, which include medicinal plants; and seek out traditional medical or religious practitioners. Various parts of medicinal plants are used: roots, bark, leaves, seeds, flowers, and fruits. 'Wild' honey is believed to have medicinal and aphrodisiac qualities (deBeer and McDermott 1989).

Other Uses of *Jaňgal* Resources

Jaňgal resources are used not only for human food and medicine, but also for animal fodder, litter for animal bedding, compost (crop production), building and household materials; are valued as cash and non-cash commodities; and are essential for ritual and ceremonial purposes (Bhattarai 1989; Diemberger 1988; Hoskins 1985; Metz 1989; Rajbhandari 1991). They are used as a source of cash to purchase food when supplies are low (Falconer 1990). The *jaňgal* also provides biological diversity and environmental services (Pimentel et al 1997).

The destruction of the forest environment affects the bulk of the indigenous population. The greatest impact is felt by women; the primary gatherers, users and managers of non-timber forest resources for subsistence purposes. Women possess a vast knowledge of forest resources (Agrawal 1986; Dankelman and Davidson 1988; Fernandes and Menon 1987; Inserra 1989; Shiva 1989). Rocheleau et al (1989) reported that women in Kenya collected several species of 'wild' food and medicinal plants for propagation and domestication in their home gardens. Continued deforestation and reduced access to forest resources (privatization of communal lands) has decreased the availability of these subsistence resources (Fernandes and Menon 1987; Golpaldas et al 1983a; and Zurick 1988).

Non-Timber Forest Products in Contemporary Nepal

One area that has received minimal attention from researchers is the role of 'minor' forest resources as an important subsistence strategy for the agricultural and pastoral population living in Nepal. These resources are viewed as easily accessible and "free" within local economies and thus their importance has been devalued. Indeed, the frequently-used phrase 'minor forest products or resources' sums up well how conventional forestry circles perceive them. "They tend to be treated as peripheral," of vital importance to local people "but not something that is a major concern for the forestry authorities" (FAO 1989b:111). The broader value of the forest (food, fiber, medicine) is lost precisely because the focus by external actors has been on the value of the individual trees as a fuel source (Falconer and Arnold 1991; Fortmann and Rocheleau 1985; Hoskins 1983; Lampietti and Dixon 1995; Sanwal 1989).

Much of the forestry policy debate has been centered on the commercial versus the subsistence value of the forests. Appasamy (1993) notes that the evaluation of forest products and services must be considered in a three-tier holistic framework. The first tier contains the ecological functions of forests, watersheds and preserving biodiversity (national parks). The second tier provides the subsistence functions – non-timber forest products used by rural and tribal communities. The third tier contains the development functions, supplying timber and wood products for the industrial sector and medicinal plants for exports. Following my research, I would add a fourth tier – cultural (including religious) functions.

The Master Plan for the Forestry Sector – produced by His Majesty's Government of Nepal (HMG), in conjunction with FINNIDA, Finnish Development Organization and Asian Development Bank – is based on information of the early 1980's and looks at the product potential of medicinal and aromatic plants, *lokta* (paper plant) and bamboo species. The major focus has been on the cultivation of medicinal herbs and aromatic plants as potential sources of income for the rural poor, and there seems to be no record of any long-term development policy for minor forest products (Malla 1982 and UNDP 1987).

Since 1990, there has been increased awareness and interest in non-timber forest products, as evidenced by several regional and country-wide seminars which discussed their important role. These include:

- The Fourth International Seminar on "Sustainable Development of Tropical Forests," held in Japan on November 17, 1990, in conjunction with the Ninth Council Session of the International Tropical Timber Association (ITTO 1991);
- The national seminar on Non-Timber Forest Products: Medicinal and Aromatic Plants held in Kathmandu, Nepal on September 11-12, 1994 (NTFPs 1994);
- The Regional Expert Consultation on Non-Wood Forest Products, Social, Economic and Cultural Dimensions, held in Bangkok, Thailand from November 28 to December 2, 1994 (NWFPs 1995a);
- The XVII Pacific Science Congress held in Honolulu, Hawaii, USA from May 27 to June 2, 1991, discussing "Society and NTFPs in Tropical Asia (Fox 1995); and
- The International Expert Consultation on Non-Wood Forest Products held in Yogyakarta, Indonesia January 17-27, 1995 (NWFPs 1995b).

The importance of natural resources for the regional economy has surpassed their subsistent value for the rural economy and the marginalized farmer. The economic value of minor forest products needs to encompass all aspects of use-value, subsistence, barter value and cash value alike (Lintu 1995b). DeBeer and McDermott (1989) assert that there have been few attempts for the valuation of non-timber forest products. A notable exception is the thorough 1993 study by Gunatilleke, Senaratne and Abeygunawardena.

Hecht et al (1988:25) asserts that small-scale extraction is more pronounced among those who are impoverished. For women, forest extraction is a major source of cash because they are often denied access to alternative means of acquiring an income in rural areas. A study in the Knuckles National Park, Sri Lanka, attempted to estimate the composite income of the peripheral communities, particularly the extraction of non-timber forest products. Results showed that although cardamom production and shifting cultivation contributed most to overall household income,[9] respectively 26

---

9 Total income includes both net monetary and non-monetary income without deducting the cost of family labor. In calculating income, farm and forest gate prices of agricultural and forest products were used. This is differentiated from cash income.

and 20 per cent, only 68 percent of households participated in cardamom production, and 86 percent of households participated in shifting cultivation. On the other hand, non-timber forest products contributed less total income – only sixteen percent – but were extracted by all of the households. A significant finding was that non-timber forest products are economically more important among the poor, because as household income increases, the contribution from non-timber forest products decreases (Gunatilake, Senaratne and Abeygunawardena 1993).

Even though these forest resources are rich in nutrients, cultural studies of farmers have neglected their importance as food sources, especially regarding household food security issues (Ellen 1986; FAO 1989a and b; Wilson 1974). Data is needed on types and species of flora and fauna harvested from the forest, the time allocated for harvesting, processing and marketing of non-timber forest products and the relationship of these activities to land tenure, proximity to the forest, family size, age group, overall income and literacy level (Gunatilleke, Gunatilleke and Abeygunawardena 1993). If forestry projects only look at the forests, and a "blind eye" is turned to the *jaṅgal*'s resources (and these "hidden" food sources), women and children who are most dependent on them will be at greater nutritional risk, and these projects will have negative impacts. The importance of *jaṅgal* resources for the overall well-being of women (health and nutrition), especially during periods of food scarcity, has been neglected. Loss of *jaṅgal* foods can lead to malnutrition, as the nutritional quality of the diet deteriorates and affects the economic and labor potential of the household.

The Mohonk Conference held in New York in 1986 declared deforestation as a major environmental issue in Nepal, and concluded that better maps of the forest resources were needed (Ives 1987). However, the mapping of *jaṅgal* resources often enhances their status and they are no longer "hidden." Outside interests arise (government or international) which can outweigh the subsistence needs of the farmers (Dove 1994). A map should include documentation of the local community's use and knowledge of *jaṅgal* resources; the relationship existing between *jaṅgal* resources and food security; use as traditional medicine for home health care, along with the religious and cultural importance of *jaṅgal* resources.

A clear understanding is needed of the current status of *jaṅgal* resources in order to protect and improve them (FAO 1989b). DeBeer and McDermott (1989:10), warn that "the development of NTFPs per se provides no guarantee that rural people receive its benefits." Thus the formalization and enforcement of the rights of traditional dwellers, and the participation of all rural communities in the control and management of the local *jaṅgal* and its *jaṅgal* resources is imperative. Only by reducing the degree of poverty can the well-being of the foraging subsistence farmer be improved and government efforts to reduce the pressure on natural resources become feasible (Ives et al 1987:332).

CHAPTER 2

The Role of *Jaṅgal* Resources

Using an ecological anthropology approach, I looked at the management of *jaṅgal* resources and their use as an adaptive strategy for household subsistence security by marginalized hill farmers. The process of change has been occurring throughout Nepal for centuries, through such means as Sanskritization, changing land tenure patterns and wage labor. One area of persistence has been the Rai's use and management of these *jaṅgal* resources. *Jaṅgal* resources have been reduced to what they are not, that is, not timber, and degraded into a set of initials NTFPs (non-timber forest products). I argue that this classification is a narrow view of the broad rich yield provided by these *jaṅgal* resources for the foraging farmer, who uses them as medicine, food, cash, construction materials, incense, etc.

By studying the flow of *jaṅgal* resources, it is possible to assess where constraints and opportunities exist, how farmers cope with them, and the ways they manage and sustain resources which are of central importance to them (and not "peripheral" resources as some outsiders believe). The broad perspective of ecological anthropology incorporates a variety of theoretical approaches, which are needed to address the research question. Because an eclectic approach is warranted, I incorporate several diverse perspectives in this *jaṅgal* resource analysis:

- ecological adaptations or maladaptations;
- ethnoecology (local knowledge systems); and
- economic anthropology (local and external economic relationships)

I have included the institution of religion to emphasize the "hidden" importance of symbolic systems, and focused on the role of women in order to look at female and ethnic status and power, household intra-variability and life cycle stages within the discussion of the total system of *jaṅgal* resource allocation and control.

The discussion of these various facets of ecological anthropology is not meant to be an exhaustive critique or a historical review of their development. Only the pertinent features, shortcomings and issues, relevant to natural resource management will be highlighted. I do not use an ecofeminist perspective, but rather will only borrow concepts that will be useful in differentiating gender aspects in the household food path of *jaṅgal* resources. These theoretical approaches have common elements: they are open, actor-centered, stress individual variability, emphasize processual decision-making, historical change and holism. A holistic approach is needed to look at whether *jaṅgal* resources are adaptive for the survival of foraging farmers.

Ecological Anthropology

Ecological anthropology has its early roots in cultural ecology and according to Sponsel (1997), cultural ecology is occasionally used as a synonym for contemporary ecological anthropology. The field of cultural ecology, developed by Julian Steward in the 1930's, used a theoretical framework for studying cultural change as an adaptation to the local environment (Steward 1955). Cultural ecology "views culture as the decisive factor" in a human population's interaction with its ecosystem (Sponsel 1987:40). Since this time, the discipline has undergone numerous changes. Cultural ecology, with its culture-environment dialectic interplay, was weak in biology (Ellen 1986; Hardesty 1977), and Rappaport (1971) remedied this by placing culture within a biological framework. This approach, referred to as an ecosystems analysis, allowed for an empirical and quantitative study of cultural phenomena. Sponsel (1997) asserts that it is this "biologization" of the ecological approach in cultural anthropology which is now labelled as ecological anthropology.

The concept of carrying capacity assumed that at least one variable was potentially limiting and researchers were obsessed with energy (White 1959; Rappaport 1968), later followed by protein (Dornstreich 1977; Harris 1977; Nietschmann 1973) which led to a carbohydrate-protein controversy (Vayda and McCay 1977). Vitamins and trace nutrients were excluded even though the nutritional quality of resources is as critical as their quantity. A single factor is not consistently limiting, since duration, seasonal changes, disasters and technology, all have an effect on environmental resources (Hardesty 1977; Bates and Plog 1991).

The systems approach has been criticized for dealing mainly with negative feedback, a synchronic approach (Harris 1965; Rappaport 1968), and neglecting processes relevant to a diachronic approach whereby systems transform themselves in response to either external or internal dynamics (Moran 1990:17). Fricke (1989:131) describes Himalayan cultural ecology's new theme as assuming a "truly processual approach that incorporates the individual and domestic dimensions of adaptation within historical and social contexts." Actor-oriented models used to study subsistence strategies and adaptation to crisis have become major fields of study (Vincent 1986). However, Guillet (1983) cautions that it is difficult to determine whether a particular strategy is more or less "adaptive" than another. Edgerton (1992) challenges the assumption that culture always has a positive function for folk populations. He warrants that an evaluative analysis must replace relativism because beliefs and practices can occur anywhere "along a continuum of adaptive value" (Edgerton 1992:206). Researchers stress the importance of studying the behavioral responses used by people in risky and uncertain situations, and the scheduling decisions where seasonal and annual resource fluctuation is substantial, and natural hazards often occur (Cashdan 1990; Vayda and McCay 1977).

Bennett (1976), Ellen (1986) and Hardesty (1977) assert that strategic human behavior and variation through time, space (boundary) and population is rarely researched. A study of time is important to account for the seasonal and other fluctuations in *jangal* resources. Space must be examined to reach an understanding of different environments, for example, secondary forested areas. Population must be looked at in terms of such things as different nutritional needs and also from the social dimension which establishes special roles in subsistence.

One social dimension is the institution of property rights where societies exert controls "on access to resources and various kinds of rules and institutional arrangements to limit exploitative activities" (Acheson 1989:376). Acheson's assertion is a rebuttal to the "open access" debate discussed in Chapter 1, and first initiated by Garrett Hardin (1968) in *The Tragedy of the Commons*, who said it is the institution of common property itself which will result in the degradation of natural resources. Common property as a social institution was illustrated in early times by Malinowski (1926). He showed that the Trobriand Islanders had a complex and variable system of rights, duties, functions and obligations, based on their common property joint ownership of large canoes. According to Bennett (1990), research is needed on the institutions which allocate and control the local population's use of resources because we need to better understand the sets of rules by which they are governed. We find that as a result of the way in which access to common property is allocated, degradation is often controlled.

Escobar (1995:105) points out that there are also institutions that are formed by outside actors that "contribute to producing and formalizing social relations, divisions of labor and cultural forms." These organizations are often headed by ruling groups and decisions and work are biased by power holders. These actors' cultural perspective of the "peasant's world" and their problems are very different from how the local people perceive them (Escobar 1995).

Reichel-Dolmatoff's (1976) study is the closest to a holistic approach of ecology. He did not use monistic determinism to explain the animal protein hypothesis because he regarded individuals in the ecosystem as biological, cultural and mental beings. Food values have a social, mental dimension, not only in terms of calories and protein, but food processing, distribution, consumption and proscriptions during critical life events. The study of a population using the household as a unit of analysis needs to consider the complex negotiations among individuals. They "embody cultural expectations, social rank, gender, hierarchies, age and demographic considerations which shape the outcomes summarized as household behavior or decisions" (Moran 1990:21). Thus a study of the social and gender relations in the production, consumption and distribution of *jangal* resources is needed.

Ethnoecology

Ethnoecology refers to how a population categorizes its environment (Levinston and Ember 1996) including the classifications, practices and techniques used in interacting with the environment (Ellen 1986). The ethnoecology approach has mainly studied native or folk classifications of environmental phenomena, for example, flora and fauna. Ellen (1986) points out that a large inventory of local knowledge of 'wild' plants may indicate the importance of gathering in otherwise cultivating societies.

A major flaw of ethnoecology has been that it rarely considers whether environmental local knowledge is adaptive or maladaptive for the community because it disregards actual behavior (Johnson 1980). Instead, ethnoecology, also known as ethnotaxonomy or ethnosystematics, "tends to consider a taxonomy of some environmental domain as an end in itself" (Sponsel 1987:33). According to deBeer and MacDermott (1989), many studies of non-timber forest products (NTFPs), for instance those reported in the ethnobotanical literature, classify NTFPs only according to the species used or products made. In Nepal, the majority of the ethnobotanical studies have focused on the medicinal properties of plants (Manandhar 1985, 1986, 1987, 1989a, 1990a, 1990b and 1990c) and ceremonial and religious plants (Bhattarai 1989).

The people in specific populations use individual management strategies to modify the environment (Moran 1990). Knowledge and experience, which varies according to local ecology and human geography, and by age, gender and class affiliations, influence these strategies (Richards 1980). Thus these bodies of knowledge held between different groups and within groups must be differentiated. The process of decision-making affecting ecological relations is still in its infancy since previously the theoretical significance of human interpretation of the environment was either treated as irrelevant or assigned to a "black box" (Ellen 1986:204).

Frake (1962) was an early pioneer in attempting to understand the actual decision-making process to account for observed settlement patterns as outcomes of resettlement rules. Conklin's (1954) work in the Philippines went beyond the classifying domain to relate it to swidden cultivation. Decision-making models (how and why people make decisions) are increasingly used in monitoring and early warning systems (Orlove 1980; Bates and Plog 1991). The perception of mountain hazards and the villagers patterns of behavioral response and traditional strategies in relation to the landscape was analyzed by Bjonness (1986) among Sherpas in the Khumbu area of Mount Everest and by S. Gurung (1989) among multi-ethnic villagers in the Kakani-Kathmandu middle mountain region. According to Ellen (1986), a complex dialectical interrelationship between scheduling activities and environmental constraints accumulates over time.

The cognized model is a description of the people's beliefs about and knowledge of the environment which may elicit unpredictable latent behavior based on socio-economic and environmental factors and relationships (Rappaport 1979). This technique is important to record changes in agricultural practices which occurred after the introduction of new technologies. Before the introduction of chemical fertilizer in Tibet in 1973, Tibetan farmers used three traditional methods to prevent insects from destroying their crops. Elde00rly Tibetans reported that all these methods were very effective. (Daniggelis 1996) :

- In the first method, a fire using yak dung was burned around their fields. Roasted barley flour, *tsampa* (T, *rtsam-pa*) was thrown in the fire so they could determine the direction the wind was blowing because smoke travels in the direction of the wind and kills the insect.[10]
- In the second method, before the farmers sowed the seed, they would either mix it with nomad salt, or the remains of rapeseed oil husks which prevented the growth of the insects.
- In the third method, the seed sowing was delayed until the start of the monsoon period based on the almanac predictions. Before the seeds were sown, the land was irrigated because insects could not grow in the damp soil. When the agricultural period was announced, all the farmers sowed their seeds at the same time.

Nazarea-Sandoval (1995:18) asserts that "the interaction between content and context (the relationship between the cognitive model and the actual circumstance in which it operates)" warrants study (see Rappaport 1979, pp.97-144). The context is the "operational reality" of the biophysical and socio-economic environment, "particularly the constraints, opportunities and strategies that define the people's resource base" (Nazarea-Sandoval 1995:14). The author cautions that these socio-

---

10 Readers may note a contradiction between the Buddhist ideal and practical necessity regarding their respect for all forms of life. In most areas of Tibet, the environment is very arid and the climate is harsh. Within these constraints, the farmers must grow enough food to feed their families, and this may require the breaking of the rule against taking life.

economic and ecological constraints vary because of the internal stratification of society according to class and gender which affects resource allocation and management. These forces are "embedded in a historical-social matrix of production" (Nazarea-Sandoval 1995:18). Nazarea-Sandoval (1995:37) studied agricultural decision-making in the Philippines among farmers as an effect of social differentiation, which is reinforced by differential access to resources and strategies as a possible cause of increasing polarization.

The cognitive method is only part of an integrated whole, and cannot be used alone (Brokensha et al 1980). The existing socio-economic institutions and physical constraints of the environment must also be addressed when understanding the foraging farmer's dilemma, which according to Agrawal (1995), shapes the processual entity of indigenous knowledge. Agrawal (1995:437) proposes *Dismantling the Divide Between Indigenous and Scientific Knowledge* (the title of his article) because "some knowledge can be classified one way or the other depending on the interests it serves, the purposes for which it is harnessed, or the manner in which it is generated." By doing this, he suggests that a dialogue initiated between the outside actors and the local people will help protect the interests of those who are disadvantaged (see Messerschmidt's 1987 discussion of the Village Dialogue). In this book, I describe the Rai and Sherpa farmer's attitude towards the *jangal* and cross-check it with a scientific approach of the nutritional value of the *jangal* foods. I describe the *jangal* from both the local actor's point of view and Western analysis (see Lett 1990).

Economic Anthropology

Economic anthropology has common threads with ecological anthropology, although their historic roots differ. The unit of analysis in economic activities related to production, distribution and consumption and exchange patterns of goods is often the household (Barlett 1989). Wilk (1996:17) pointed out, however, that economic relations within the household have been neglected and even concealed. The study of resource flow needs to look at the dynamic interactions within the multi-dimensional social system, from the household, to the local, regional, national and international level. Durrenberger (1996:367) cautions that "economic systems are complex." The complexity arises from the influence of microstructures, tenure patterns, local labor markets, kinship and social networks, rituals and household decision-making strategies.

Economic anthropologists are concerned with the role of risk (unpredictable variation in environment and economic conditions) and uncertainty (lack of information) affecting the production system of farmers as well as the coping strategies used by the farmer (Cashdan 1990). Ortiz (1990) says that Barlett's (1982) suggestion that the poor peasant farmer places a higher priority on reducing risks than on amassing profits, is based on a wide range of past experiences. Escobar (1995:158) points out that "peasants behave rationally; given their constraints, they optimized their options, minimized risks and utilized resources efficiently." Under conditions of uncertainty, decisions are likely to be flexible and leave room for a number of contingency strategies (Ortiz 1990).

Plattner (1989) stresses that the analysis of the strategies of subsistence and exploitation of natural resources must address social stratification, gender and common property because they are important in looking at economic stratification and differential access to resources. Although peasant villages are often seen as uniformly "poor" by many outsiders, there exists an internal wealth and status difference within these communities (Barlett 1980). The distribution of development costs and benefits between different sectors of society, and between different strata in a farming community is a central issue in development planning (Barlett 1980). External forces such as governmental structures, middlemen and markets, national parks and the impact of tourism, and outside employment opportunities that may increase the pressure on the *jaṅgal* environment must be included in the overall analysis of the subsistence situation of the foraging farmer.

Religion

The term environment in the "Geertzian" sense focuses on how cultural symbols help to explain the actor's interaction with the environment (Ortner 1978). The symbolic approach places culture within the ideational realm, that is, cosmological beliefs and religious ritual. Geertz (1965:167) interprets religion as a symbolic system and demonstrates that "sacred symbols deal with bafflement, pain and moral paradox by synthesizing a people's ethos and their world view." Ortner (1984:130) remarked that although Geertz's model was actor-centered, "he did not develop a theory of action or practice."

Tucker discusses how different religious worldviews have been inherent in forming moral attitudes toward nature with both positive and negative results (when humans are seen apart from nature) (Tucker 1997). Sponsel (forthcoming) suggests that the recent trend is to view religious practice as a positive approach to developing sustainable environmental ethics. Anderson (1996:167) points out that "in traditional societies, religion, worldview, and resource management strategies are often inseparable," encouraging resource conservation and an emotional respect for both plants and animals. Tucker (1997:21) attributes this to the fact that"the consciousness of the spirit presence within nature rather than simply beyond nature is particularly high in these traditions."

Dove and Rao (1990) assert that "one of the few remaining traditional institutions with any power of 'sanction' is religion." Religion acted as a form of social control in Tibet prior to 1959. The Tibetans believed if the community lived in harmony, there would be less hail, but if there was disharmony, then God sends more hail. Natural hazards are a frequent occurrence on the Tibetan high plateau. Hail storms can destroy an entire agricultural crop. The main function of a *ngakpa* (T, *sngags-pa*), a Tantric practitioner, was to stop hail before it could damage the crops (see French 1995: pp.197-203). In ancient times, the hailstorm was seen as an act of revenge. Milarepa (a renowned saint and poet 1038-1122 AD) had the power to cause hailstorms to destroy crops and punish his enemies (Paul 1982). Reinhard (1987) describes the anti-hail ritual performed by the Helambu in North-Central Nepal. In my own research (see Chapter 6) the *pūjā tos* is performed by the Rais to prevent hail.

In the Mahāyāna tradition of Buddhism, and that practiced by the Sherpas, there is an inherent respect for nature. Positive acts towards nature are one means to achieve good *karma* and a person's chance for enlightenment (French 1995). Environmental ethics inherent in Buddhism "advocates reverence and compassion toward all life, including invertebrate and vertebrate animals" which promotes biodiversity and conservation (Sponsel and Natdecha-Sponsel 1993:83). The authors further add that a long history of mutualism exists between Buddhism and trees and forests. This relationship has contributed to the dispersal and maintenance of particular plant species (Natadecha 1988).

Rappaport (1979:174) was important in initiating the orientation that ritual may serve an important functional role for the environment. Ritual performances recall and present in symbol form the underlying order that guides the social community (Leach 1976). The ritual is a prototype of the way the people's lives are ordered because they want to appease the Gods and maintain order.

"Myth and ritual are particular contexts of social life where the tie with the past and the cosmos is renewed (and with this the existing social relations). They have their own plurality which is played out within the political sphere in the community. In a community where different traditions coexist, the choice is basically a question of political (and religious) consensus" (Diemberger 1992:CH8, pp.4-5). "Rituals and myths which interpret history and geographical elements" can create a common identity in a community, express social cohesion and legitimate political power (Diemberger 1992:CH4,p.33).

Animate and inanimate objects have religious importance for cultures throughout the world. Parkin (1992:73) discusses the totem objects (usually flora or fauna) for specific clans among Central India's tribal groups as protecting that particular resource. Totems, featured in the clan's origin myths, serve as emblems distinguishing individual clans and descent groups. The totem object's protection is generally "clan specific" and the belief is held that "any serious decrease in the species will threaten the existence of the clan." Jackson (1993:668) cautions that totemism may support 'speciesism'[11] and that a material analysis into the element protected and the major beneficiaries of this act is warranted.

Religious beliefs and cultural life have been intricately intertwined with the *jańgal*. Sacred forests (groves) are perceived by villagers to be the home of spirits and deities and are a form of environmental protection at the local level. Examples of sacred groves have been discussed by Cronin (1979), S.Gurung (1989), Mansberger (1991) and Pei (1985) which will be presented in further detail in Chapter 9.

The Mundari people of Assam regularly honor the deities of the sacred groves at agricultural festivals to ensure abundant crops (Altman 1994). Altman explains that a tree becomes sacred because the power that it expresses is recognized by the culture. Altman (1994) and Kabilsingh (1996) uses the *pīpal* or *bodhi* tree (*Ficus religiosa*), the most sacred tree to both Hindus and Buddhists, as an example. For the former, it is the home of Krishna, Brahma, Vishnu and Shiva, and for the latter, it is associated with Buddha's achieved enlightenment. Among the Newars (a Tibeto-Burman group which are the original inhabitants of Kathmandu Valley), young girls are married to the fruit of the English Bael tree (*Aegle marmelos*) to symbolize permanence. Although the girl is free to remarry, she remains wedded to the Bael fruit for life. If her human husband should die, she does not become a widow or an outcaste (Altman 1994; Majupuria and Joshi 1989). For the Sherpas, certain trees, bushes, cliffs, water or mountains are abodes of deities. These resource species also have great economic value for them and the species on which they depended the most were given a sacred status and were preserved (Furer-Haimendorf 1964; Stevens 1993).

Bernbaum (1992:xiii) discussed how "people have traditionally revered mountains as places of sacred power and spiritual attainment." Buddhism is one of five religions that consider the Himalaya sacred. The mountains have great value for the Sherpas because of the residing mountain deities who are perceived as guardians of the land and all living beings. In the Khumbu region, a local god is associated with the sacred mountain Khumbila. Sherpas refuse to climb this mountain out of fear of angering the deity and causing harm to the nearby villagers. Before climbing Mount Everest, referred to as Chomolangma by Sherpas and Tibetans, they make offerings and burn incense to the deity as a prophylaxis to prevent disaster. If accidents occur, some locals believe it was because the residing deity had been angered.

---

11 Speciesism is a key concept in ecocentric discourse and has been defined as 'a prejudice or attitude of bias toward the interests of members of one's own species and against those of other species' (Ryder 1974 quoted in Eckersley, 1992:43).

The Role of Women in Environmentalism

Women are highlighted in this research because of their vital role in the environment and "food path"[12] of *jaṅgal* resources. Agarwal (1991) asserts that poor peasant and tribal women maintain a reciprocal link with the natural resources. Women possess an intergenerational cultural knowledge system of the *jaṅgal* and make economic, medicinal and nutritional use of *jaṅgal* resources, including decisions regarding its management (Rodda 1991). Escobar (1995) and Rodda (1991) argue that women are viewed as "invisible farmers" and because of this, their local knowledge has been ignored and development has had a detrimental effect on their economic position and status which has increased their work load. The authors further stress that the socio-economic processes associated with poverty are deeply intertwined with environmental degradation. These women lack access to appropriate technology, inputs and credits, lack the right to land and property and educational opportunities. A contrasting view of the synergy between gender interests and environmental conservation has been presented by Jackson (1993:649) who argues for "a broader gender analysis to look at women in relation to men, the disaggregation of the category of 'women' and an understanding of gender roles as socially and historically constructed, materially grounded and continually reformulated."

Nutritional anthropology is included here because it looks at how individuals, especially women, organize, locally acquire, distribute and consume *jaṅgal* resources. The cultural categories used by women and other village members in identifying *jaṅgal* ingestibles were inventoried. Adaptation and maladaptation of varying degrees can occur when traditional societies are confronted with a changing environment or having to cope with an unfamiliar environment (Dewey 1979). Rai and Sherpa ethnic groups living on the "fringes" of a *jaṅgal* were studied to look at how and why the *jaṅgal* is important to them.

Women's role in environmentalism intersects with economic anthropology because both are concerned with "power relations" and "inequity" within the household (Barlett 1989). Culture change, specifically, Sanskritization (a desire to achieve a higher social status which entails, among other things, prohibiting particular *jaṅgal* ingestibles) may negatively affect the health and nutritional status of vulnerable groups, women of child bearing age and children, birth to five years. When a group is becoming Sanskritized, the bias in distribution of food within the household toward males would be more marked (Agarwal 1986; Bennett 1988; Dettwyler 1989; UNICEF 1987).

---

12 The food path is the flow of *jaṅgal* ingestibles from the gathering to the processing, distribution and consumption by household members.

The interactions of environment and culture influence opportunities and constraints for dietary strategies at the individual and household level (Grivetti 1981). Historically cultural anthropologists have paid minimal attention to the nutritional, social and cultural aspects of the food systems of the communities they have studied. Ethnographers usually described food procurement within the category of "economics," but food preparation, dietary patterns, and eating habits had received less systematic treatment. Cultural studies of farmers have only mentioned 'wild' resources in passing in their ethnographies (Ellen 1986). An assessment of the nutritional **quality** [emphasis mine] and quantity of food and dietary patterns is rarely a subject of general ethnographic research (Pelto and Jerome 1978). There are exceptions, such as Audrey Richard's (1939) pioneering work on the Zambian Bemba; and also those by Bennett (1946), Mead (1945), Gross and Underwood (1971), and Pollack (1992).

An important role of ecological anthropology is to look at contemporary problems of the local people. Bates and Plog (1991) assert that a major contribution of ecological anthropologists is to explain the local group's coping strategies in order to assist development planners to either enhance them or at least to avoid interfering with them where possible. At the Mohonk Mountain Conference held in New York in 1986, a supra-national region-wide strategy was formulated to look at problems of population, resources, and the environment using a catalogue of different cultural and sub-cultural frameworks (Ives et al 1987:343) Non-timber forest products were identified as a research priority in The Makalu-Barun National Park and Conservation Area Management Plan (Shrestha et al 1990:58-59).

My research looks at how cultural, political and historical factors affected the hill farmer's interaction with his/her environment, and examines nutrition and foraging in subsistence agricultural and pastoral societies as adaptive strategies. These strategies are used by the farmers in response to the availability of *jaṅgal* resources which are the basis for choice when faced with issues of household security. This research provides baseline data on patterns of *jaṅgal* resource use under present environmental conditions (the biological, technological and cultural/social environmental context) and discusses historical land-tenure relationships and management in the area. These findings are important because development projects invariably lead to social changes that may affect the knowledge, use and management of the *jaṅgal* environment and in so doing may adversely affect the lives of the rural population. Resources regarded as 'wild' by outsiders (that is, high-caste Hindus and external agencies) is of central importance to the foraging farmer whose main survival strategy is dependent on them.

CHAPTER 3

The Research Area

Nepal is a small country encompassing an area of 147,181 square kilometers. The resource-poor mountainous country is wedged between India and Tibet. In 1991, Nepal's population consisted of 18.5 million people with a growth rate of 2.1 percent. In the 1993 United Nations Human Development Report (UNHDR), Nepal is included among the world's least developed and poorest countries ranking 152 out of a total of 173 nations. The UNHDR is a new approach used to compute a nations' economy and health based on social and economic indicators rather than per capita income, gross national product and trade balance. The life expectancy of the population is estimated at 52 years, one of the lowest in Asia. The life expectancy figure for women is even lower than 52 years. In the 1991 census, the literacy rate was recorded at 39.6 percent, and women account for only one-third of this figure (Karan and Ishii 1994). More than one-half of the children, less than five years of age, suffer from moderate to severe malnutrition (Savada 1993:100).

Nepal has a low per capita income because the majority of the population is dependent on subsistence agriculture. Transportation and communication networks still rely on runners and porters in many areas which intensifies the isolation of most of the country and its rural population. Export trade involves a small number of primary products with little or no processing. Nepal is characterized by a heavy dependence on foreign trade (tourism), a high population growth, unbalanced regional development and low productivity of agriculture. Nepal's major hindrance in having an economic policy independent of India is that the use of Indian ports must be based on common agreements. In 1989, when treaties with India expired, seaports were closed which led to a scarcity in kerosene and foodstuffs in Nepal. Trees were cut by the Nepalese people as a means to alleviate the fuel shortage crisis (Karan and Ishii 1994).

Nepal's early history under the monarchy regime was discussed in Chapter 1. In April 1990, several political parties launched a movement to demand democracy based on a multi-party system. At this time the *Panchāyat* system was disbanded.

Nepal is divided into three ecological zones, the Tarai, the Middle Hills and the High Himalaya, with peaks greater than 8,000 meters.

There is a vast geographic, cultural and ethnic diversity existing in Nepal. More than seventy-five major ethnic groups are present throughout the country. The majority of the population are Hindu of Indo-Aryan origin who live mainly in the Tarai and in the lower valleys of the Middle Hills. Groups of Tibeto-Burman origin have inhabited mainly the upper middle hills and mountain region. The Indo-Aryan groups have a caste hierarchy based on purity and impurity with Brāhmans at the top, the Tibeto-Burman groups in the middle and the untouchable castes at the bottom. The Indo-Aryan group is the predominant and official culture of modern day Nepal. Caste traditionally determines occupation and the pattern has been for the highest castes (Brāhman and Chhetri) to dominate government, business and politics (Blaikie et al 1980).

The Arun Valley

The Arun Valley has one of the world's deepest river gorges and contains one of Nepal's most biodiverse ecosystems. The vegetative description of the Arun Basin will be discussed in Chapter 5. In the Nepalese Himalaya, the Arun Basin is one of the wettest basins with more than 4,000 millimeters of annual rainfall in certain regions (including the Apsuwa Valley). The southwest monsoon occurs from June to October with pre-monsoon thunderstorms beginning in April or May. This is a rugged area and the thrust planes and cold climate produce effects of shearing in the rocks making them susceptible to erosion and instability causing rock slides (Shrestha 1989).

Historically marginalized hill farmers, particularly the Rais, have foraged for *jaṅgal* resources as a survival strategy. As agricultural land loses fertility and becomes increasingly reduced, poorer farmers are even more dependent on the nutritional, economic, social and cultural aspects of the *jaṅgal* and its resources.

An isolated research area was chosen because it was assumed the indigenous cultures will have retained more traditional social structures and economic adaptations because, on the one hand, the environment is likely to be more intact and because fewer opportunities for wage labor, cash crop sales, etc. exist. In the Apsuwa Valley, as well as throughout the Sankhuwasabha District, villagers experience periodic crop failure and seasonal migration occurs in a majority of households. During these crisis periods, it is expected that people will increase their dependence on *jaṅgal* resources. Because the area has vast biodiversity, it is probable that the *jaṅgal* foods will be gathered to buffer the periods of crop failure and inadequate produce from livestock.

The research area, which includes the villages of Gongtala, Dobatak, Saisima and Yangden, is located in Ward 9,[13] the upper northwest Apsuwa Valley of northern Sankhuwasabha District, Eastern Nepal (see Figure 2). This area is remote because it is not on the well-trekked Makalu Base Camp trail. From Hile, one of the major market and trading towns in Eastern Nepal, it is a seven to eight day walk to reach the area. An alternative route is to fly to Tumlingtar and then walk uphill for three days through extremely rugged terrain to reach the isolated area.

These villages face southeast and are on slopes where cultivation is possible. The Apsuwa *Kholā* (river) is a perennial western tributary of the Arun River, *Phu Chu* (T, *phu-chu*), which originates on the Tibetan side of the border (Shrestha 1989). Mt Chamlang, 7,319 meters high, situated between Mt. Everest on the west and Mt. Makalu on the east, can be seen from the research area. The study area covers an altitude from 1,000 meters (Yangden) up to the *lekh* more than 4,200 meters high. There are no roads and the local people have to rely on treacherous narrow footpaths which become very slippery, and especially dangerous, during the monsoon season and winter months. Locally-made bridges are unsafe and rain swollen rivers often wash them away during the monsoon . Natural hazards, for example, horrendous rainstorms, are an ever present problem and landslides frequently occur dislocating families from their land.

Figure 2: The research area, including the villages of Gongtala, Dobatak, Saisima and Yangden

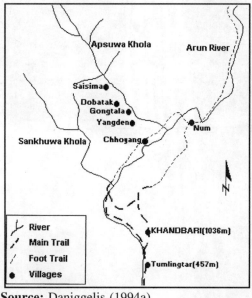

Source: Daniggelis (1994a)

---

13 A ward has approximately 800-1,000 people and is the smallest geographical and political division in the country.

There is no electricity and indoors, the fire or kerosene lamp suffices; while outdoors, a flashlight, a flaming bamboo torch or the moon's illuminance improves night visibility. The Singh Devini Primary School, Grades 1-5, is located in the village of Yangden. It was built in 1976 by the villagers using materials from the *jaṅgal* and family savings. The first school teacher was Purna Bahadur Rai, the former *Pradhān Panchāyat* for the area. Tamkhu, a distance of a one-day walk, has a school up to Grade 10 and students board there from Dobatak and Saisima. The nearest Western-style health post is located in Tamkhu. Piped water was recently established in the Sherpa area, but does not exist for the Rais.

Ephrosine Daniggelis

A log bridge over the Apsuwa *Kholoā*

Twenty months of research (between June 1991 to March 1993, and April 1994) was conducted among two ethnic communities, the Khumbu Sherpa and Kulunge Rai, residing in the upper Apsuwa Valley of Eastern Nepal. These populations live within the Makalu-Barun National Park and Conservation Area in northern Sankhuwasabha District, formally established in December 1992. They were studied to understand differences in their knowledge, use and management of *jangal* resources, and look at their responses to problems of household food security. The Makalu-Barun Conservation Project (MBCP) is a joint endeavor by the Department of National Parks and Wildlife Conservation of His Majesty's Government of Nepal in conjunction with The Mountain Institute, an international non-governmental organization. Makalu-Barun National Park and Conservation Area is one of fourteen protected areas, including eight national parks, in Nepal. The goal of the Makalu-Barun Conservation Project is to protect the wealth of biological diversity of the Makalu-Barun area, which contains some of the last remaining virgin forests and alpine meadows in Nepal and promote community development for the people living in the conservation area. Non-timber forest products are one of several areas of focus within their management plan for the park (Shrestha et al 1990).

Saisima is the only permanently inhabited village situated within the national park boundary, Nepal's first Strict Nature Reserve. The villages of Gongtala, Dobatak and Yangden lie within the Conservation Area, an area of 830 km2 designed to encourage the 32,000 local inhabitants to become actively involved in biodiversity protection, cultural conservation and joint parks/people management systems (Shrestha et al 1990).

Four Village Histories

The Sherpas trace their ancestry to Tibetans from Kham, Eastern Tibet, who emigrated about 450 years ago to the Khumbu region near Mt. Everest (Oppitz 1973). An elderly Sherpa told a story of why the first Sherpa settlers came to the area now containing Dobatak and Saisima villages.

About 1,185 years ago, Guru Rinpoche (Padmasambhava) came to the area in the Apsuwa Valley which is now known as Beyul Khenbalung and stayed for one week. Guru Rinpoche hid many things in the valley because he felt that there would be trouble in the world. We came to this area after we read this story in the book of Nèshè (gnes bshed). One day the world will be destroyed and only in Beyul Khenbalung can people live and be alive in the next world. There are other beyuls in the world but only the beyul of Khenbalung is located here.

In Tibetan mythology, Khenbalung holds the possible entrance into the hidden mystical valley of Shambhala (Bernbaun 1980). Padmasambhava, an Indian Buddhist yogi, founded Tibetan Buddhism during the 8th Century A.D. Padmasambhava, often referred to as Guru Rinpoche by Sherpas, is the central "saint" of Sherpa Buddhism (March 1987:379,14f).

In 1993, the population of Gongtala, Dobatak and Saisima consisted of approximately 135 Khumbu Sherpas living on the west ridge of the Apsuwa *Kholā*. Khumbu Sherpas are agropastoralists cultivating dry highland crops on steep slopes. They are heavily dependent on animal husbandry which helps meet food needs. Large amounts of grains are required for making *chang* (T, *chang*, a fermented grain beer). The water sources are small rivers located near the villages. Swidden agriculture is more apparent near the villages of Dobatak and Yangden than at Saisima. Near Saisima, one of the forests, along with the wildlife, is protected by religious sanctions. Hunting, grazing of livestock and the collection of firewood and *mālingo* (*Arundinaria aristatla*) are prohibited.

Saisima

Saisima is 2,300 meters (7,550 feet) in elevation and a four-hour walk northwest of Gongtala. In upper Saisima, there is the Denchen Chöling *Gönpa* (T. *dgon pa,* a Buddhist monastery) headed by Ngawang Thaye Lama, and six small houses inhabited by nuns. The lower village, a ten-minute walk down a footpath, consists of 25 Sherpas living in nuclear families in four bamboo houses.

Saisima is the oldest of the four settlements in the research area. Ngawang Thaye Lama discussed the early migration of his maternal uncle's generation from Pangboche, a village in the Khumbu Region close to the present Everest Base Camp, about five generations[14] or 130 years ago, and told a story of the early settlers.

> *About 1920, there was a fight among the villagers of Saisima and a Sherpa named Gyaltsen*
> *was killed. The people fled because they feared they would be severely punished by the*
> *jimmāwāl (the local tax collector who also acted as a policeman to maintain law and order).*
> *The Sherpas fled to various destinations: Chitre (a Sherpa village two day's walk to the*
> *west), Darjeeling (Northeast India), Walung (a one-day walk to the northeast) and to the*
> *Solu region. Only the stones of the original houses remain in Saisima.*

In 1974, a group of Sherpas came from Memerku, Chheskam, in the Solu region to build a Buddhist *gönpa* in the area of Saisima because the legendary Khenbalung is located nearby. Each of the four households paid Rs 2,000[15] for the land which they bought from a Sherpa living in Dobatak.

---

14 According to Fricke (1984), one generation is equivalent to 26 years.
15 In 1991, 42 *rupiyā* was equivalent to US $1.

Table 1 shows the demographic household structure for Saisima village.

Table 1

Gender distribution and age grouping for the four households[16] in
lower Saisima village[17]

A g e	Males	Females	Total	%
0 - 5 years	2	2	4	16
6 - 14 years	6	4	10	40
15 - 45 years	3	6	9	36
46 - 64 years	1	1	2	8
65 & > years	–	–	–	–
Total:	**1 2**	**1 3**	**2 5**	**1 0 0**

The lower Saisima village consists of 25 Sherpas who live in four nuclear family households with both parents present and with 3, 4, 4 and 6 children respectively. Two of the household heads do seasonal trekking. Three household heads are of the Thaktu lineage from the Hongu Valley and one house is of the Chayaba lineage.

The Sherpas practice Nyingmapa or Red Hat Buddhism, the oldest sect of Tibetan Buddhism. All the Sherpa villages have *lamas* (*T, bla·ma,* Buddhist priests) who are responsible for conducting ancestor rites, curing illnesses, and other ritual obligations. In the Nyingmapa sect, there are no restrictions regarding marriage for *lamas* practitioners. The *gönpa* in Saisima was the only *gönpa* in the Yaphu Village Development-ment Committee until Chitre, a village, a two-day walk to the west, built one in 1994.

The present *gönpa* is relatively new, built in 1986 by the community of Saisima, *lamas*, nuns and lay people. Before the *gönpa* was built, the religious practitioners who were non-residents of Saisima, stayed in temporary lean-to shelters and studied at the *lama's* home. Nawang Thaye Lama became the head *lama* after the previous senior *lama* died in 1979. A 24 year-old Sherpa man, whose parents died when he was young, lives with the *lama*. There are eight nuns who farm and grow their own food. Many nuns and *lamas* in this area go to Thupten Chöling *gönpa,* Junbesi, Solu region, for special teachings which last for one to two months. The expenses incurred for food and housing at the *gönpa* can cost about Rs 2,000. All of the *lamas* study with Nawang Thaye Lama in the new *gönpa* and five times a month, a prayer gathering is held there.

---

16 In this research, a household is defined as the basic unit of production, distribution and consumption for the landholding and farming unit eating from a common cooking hearth.

17 The demographic household structure was broken into these respective age groupings because I was interested in the vulnerable groups, that is, children less than five years of age and women of child bearing age, 15-45 years. Also, because of the small sample size, I did not use five year age increments because it would not be statistically significant.

Becoming a nun can be either by individual choice or family decision. One young woman in Gongtala confided in me that she had wanted to become a nun but because her family needed herders for their high-altitude animals, she had to abide by her parent's wishes and follow a secular life even though her father and two of her brothers are *lamas*. Another nun during meditation at the *gönpa* became romantically involved with a young *lama* from Dobatak much to her parents dismay and left the monastic life.

Dobatak

Dobatak, a village of five households, is about 1,980 meters in elevation and a one-hour walk northwest of Gongtala. Dobatak was settled when Sherpas arrived from Khumjung Khunde (3,750 meters) in the Khumbu region about four generations ago. They chose to settle in this area because Beyul Khenbalung is located here. The first Sherpa to settle in this area had seven grandsons who now own land in Dobatak and Gongtala. Table 2 shows the demographic household structure for the village of Dobatak.

Table 2

Gender distribution by age group for the five households in Dobatak

Age	Males	Females	Total	%
0 - 5 years	3	-	3	9
6 - 14 years	3	3	6	17
15 - 45 years	8	11	19	54
46 - 64 years	2	3	5	14
65 & > years	1	1	2	6
Total :	17	18	35	100

The village of Dobatak consists of 35 Khumbu Sherpas, an average of seven persons per household unit. Dobatak has one nuclear family household; one incomplete nuclear family headed by a widow and three extended families consisting of parents with married sons and their children. One Sherpa household includes a Kulunge Rai woman to help fulfill household labor needs. Everyone talked of the Rai as a member of the Sherpa household.

The widow who headed the incomplete family has three non-resident children still tied economically to her. She receives remittances from two of her sons who live in Kathmandu where they operate a restaurant. Another son lives in America, sponsored by people that he met while working as a guide on a trek. Several children in the village attend the school in Tamkhu and board with relatives living there. During their vacation, the children return to Dobatak to help their families with agricultural activities.

Gongtala

The village of Gongtala, consists of a cluster of fourteen houses situated on a slope at an elevation of approximately 2,250 meters. Gongtala, referred to in Nepali as *Maghan*, was first settled in 1950 by Sherpas of the Pinasa clan who had migrated from Shipding, a Rai village north of Tamkhu. Their ancestors had previously lived in Solu Khumbu. At that time, there were only three households, the informant's, another Sherpa household and a Newar household. A Sherpa family arrived ten years later from Dobatak.

The household developmental cycle is shaped by the culturally-recognized life transactions: childbirth, marriage, inheritance and death (Fricke 1984). The structure of the household is very complex and involves people who are often not residential household members but work outside of the village and may send remittances. It is also common for a household to have several members who attend school elsewhere. School fees are required but normally the students board free with kin. Similarly the participation of family members as *lamas* and nuns provides ties to the wider Buddhist world for rural households. Because of the unusual nature of Buddhism in this area, *lamas* and nuns do not reside in monasteries and are not supported by the local community. Normally they farm for themselves but at times they may draw on family funds. Table 3 shows the demographic household structure for the village of Gongtala.

Table 3

Gender distribution by age group in Gongtala based on eleven households

Age	Males	Females	Total	%
< 5 years	9	8	17	26
6- 14 years	5	10	15	23
15- 45 years	13	16	29	45
46- 64 years	2	2	4	6
65 & > years	-	-	-	-
Total:	**29**	**36**	**65**	**100**

The village of Gongtala has an average of six persons per household unit. The eleven households comprise: seven nuclear families (biological parents with children); one incomplete nuclear family, female-headed by a widow; two augmented nuclear families, an older married brother and family with a younger sister, and a nephew whose father lives in Kathmandu resides with his uncle and family; and one form of joint family consisting of two married brothers with their families. Six women for periodic intervals become *de facto* household heads when their husbands seasonally migrate for trekking purposes. Gongtala has five *lamas*.

Only eleven of the fourteen households were surveyed. Two land-owning families do not meet the criteria of a 'household' of Gongtala because the members live most of the time in Tamkhu. During the research period, one household, which had two brothers, split forming two separate households and land holdings. This accounts for the addition of a fourteenth household.

The Himalayan area is known for its physical mobility. Pilgrimages, sometimes lasting several years, are a normal part of life. The early economic system valued both pastoral movement and long distance trade. An examination of several households reflects the persistence of traditional, as well as modern patterns of migration. Transhumant herding practices common in this area are an example of an ongoing traditional pattern. Also, the *lamas* and nuns in this area go on pilgrimages to Thupten Chöling *gönpa* in Junbesi for one month of religious studies. Modern patterns of movement are for trekking purposes, a recent phenomenon in this remote area. Nowadays, many of the younger generation are settling in urban areas, such as Kathmandu, opening trekking agencies and restaurants.

The face of Gongtala changed during my field stay. In June 1991, when I first arrived, there were only three houses made of stone with wooden roofs. My house also became part of the traditional landscape. However, within a period of two years, five traditional houses were re-built of stone, with wood planks, a sign of wealth and prestige. Money earned while trekking and/or borrowed from relatives residing in Kathmandu helped to defray the expenses incurred to build these houses. Although the houses were very expensive to build, the price varied according to the laborer's ethnicity and the agreed arrangements. In 1993, one Sherpa paid another Sherpa Rs 7,500 for his labor in building the house. Another Sherpa paid a Kami, a blacksmith and one of the lowest castes of untouchables, his *mit,*[18] Rs 1,700 for two months of work to build his house. This averaged out to Rs 28 per day, almost one-half the usual wage labor rate of Rs 50 per day.

By March 1993, only four of the poorest households (and the two uninhabited houses) still retained the traditional style of Sherpa dwelling. These houses are made of various types of bamboo and all roofs are made of mats woven from *māliṅgo* (*Arundinaria aristatla*). Each house has a loft to store bamboo implements and dried grains. A log ladder with carved steps is used to reach the loft.

The villagers in Gongtala belong to two patrilineal clans: Pinasa and Salaka. Four households are Pinasa and seven households are Salaka. The Sherpa inheritance rule allows each son an equal share of the natal family patrimony, including land and

---

18 The practice of *mit*, a ritual friendship formed between two males or two females, is discussed in Chapter 4.

Author's house in Gongtala

movable property. These rights are normally transferred to the elder sons after a few years of marriage. Initially, patrilocal residence occurs until the sons set up their own independent nuclear family. The youngest son, in addition to receiving the parental house, is theoretically obliged to remain with his parents to feed and care for them out of this last share of their estate. His bride is brought into the natal household forming an extended stem family (Ortner 1978:46).

Yangden

Yangden is at a lower altitude, about a one-hour walk downhill southeast of Gongtala and at the upper limit of the sub-tropical zone. Yangden consists of thirty-two Kulunge Rai households which are spread over the landscape. The highest is located at 1,800 meters, and three houses are situated near the Apsuwa *Kholā* at 1,000 to 1,300 meters.

An 82-year-old man, the oldest person in the entire research area, discussed the history of Yangden and the surrounding area. His grandfather's great-grandfather was the first to settle in this area.

The Khambu Rai came to Mahakulung about thirty-five generations ago. Our ancestors, the Kulunge Rai came to Yangden about seven generations ago. The Kulunge Rais came from Mahakulung, Solu District west, located on the way to Solu Khumbu after Chitre. The Kulunge Rai came to this area now known as Yangden because they were told it would be an easy place to live. The farmers found a lot of jaṅgal and the soil was fertile. At that time, the maize was good and grew well. The old generation felt it would be easy to spend a life here. Ban tarul[19] and ghar tarul[20] were plentiful in the jaṅgal. Food was available here. In Mahakulung, food was scarce and this is the reason why the people left. The Sherpas came sixty years ago and gave one or two rupiyā for the land.

Khambu is the term often used to refer collectively to the Rais. The entire area which includes the Hongu Valley and adjacent areas were included in the Kulunge tribal *kipat* territory (communal land), known as Maha (greater) Kulung (McDougal 1979).

Today, it takes about two hours to walk the entire distance of Yangden. All of the one-story houses are made of split bamboo matting for both the walls and roof. Agricultural implements, harvested grains and seeds are stored in the loft. A veranda is attached to the front of the house. Around the house are usually found banana trees, chicken houses and pigs in a fenced enclosure. The water sources are rivers, though one family reported having a well. Depending on the household's locale, it can take from one-half to three hours to collect the water.

The basic subsistence strategy for the Rais is mixed agriculture, supplemented with animal husbandry. The lower elevation and warmer climate of Yangden, compared to Gongtala, allows for a greater diversity of cultivated crops which supply an important means of bartering with the Sherpas. Vegetables are grown in gardens close to the house. Seventeen houses have at least one fruit tree, banana and/or peach.

---

19 Ban tarul is *Dioscorea versicolor*.
20 *Ghar tarul* is *Dioscorea alata* L. (greater yam).

A Rai household in Yangden

The Kulunge Rais are 'Kirātis,' one of the many Tibeto-Burman speaking peoples (McDougal 1979:3). The Rais, and, some authors say, the Limbus are thought to be the descendents of the ancient Kirātis (Caplan 1970). However, according to Vansittart (1915:7), the Rais are the true descendants of the Kirātis, but because of intermarriage between the two groups, the Limbus are also included within the 'Kirāti group'. "Kulung" in its broadest sense can be used for all Rai groups that are indigenous in and near "Mahakulung," the region between the Hongu River in the west and the Sankhuwa River in the East (Hanbon 1991). Rai is technically a term meaning 'headman,' or 'chief.' When Prithivingrayan Shah captured Majh Kirant (Middle Kirant) during the Gurkha Conquest, he bestowed power upon local men to act as intermediaries between the people and the state to rule their territory (Regmi 1978). Over the years, Rai has become the popular generic term of reference for this entire ethnic group (Bista 1967:32).

A Yangden Kulunge Rai

Table 4 shows the demographic household structure for Yangden.

Table 4

Gender distribution by age group in Yangden for 32 households

Age	Males	Females	Total	%
0 - 5 years	11	15	26	15
6 - 14 years	26	24	50	29
15 - 45 years	39	35	74	43
46 - 64 years	7	5	12	7
65 & > years	4	6	10	6
Total :	**87**	**85**	**172**	**100**

Yangden village consists of 32 households having a total population of 172 Kulunge Rais, an average of 5.4 persons per unit.

The type of family structures explain the relationship of kin networks and residential units, and the mobility and types of family arrangements. In Yangden, there are a variety of kinship residential relationships as shown in Table 5.

Table 5
Kinship residential relationships in Yangden

Kinship residential relationships	Number (type)
Extended families	9
Incomplete extended family	1 (fa's fa & son's d)
Biological nuclear families	8
Augmented nuclear families	3 (consisting of: a fz, a step d and a ybc)
Incomplete nuclear	3 female-headed
Incomplete nuclear	2 childless
Single person households	2 female-headed, with one HH semi-dependent living alone
Incomplete polygynous nuclear	1 (wife went to Sikkim)
Sororal polygynous families	2
Non-sororal polygynous family	1

Note: fa's fa= father's father, son's d= son's daughter, fz= father's sister, step d= stepdaughter, ybc= younger brother's child, HH= household.

There are five-female headed households: in two of the households, the husbands had gone to Sikkim (one with his senior wife); one, a woman whose husband had died; one woman had never married; and one elderly widow lived alone. One elderly couple recently moved to Yangden from across the Apsuwa *Kholā* after a landslide forced them to flee from their home. They lost all their land and now live on land of kinsmen in Yangden.

In Yangden, there are five polygynous families. In two of the households, the wives are two sisters and in another, one man has three unrelated wives. In several of the polygynous arrangements, the Rai and his two wives sleep in one room. The Sherpas joke that the Rai needs two wives, one to keep him warm on each side. One Rai whose first wife is 32 years old said he took a second wife (his first wife's younger sister who is only 18 years old) because he wanted a son. He has three daughters. The second wife, however, gave birth to a baby girl.

Cash income mainly comes from selling woven bamboo products to Gongtala and neighboring Rai villages. One Rai man earns money from occasional trekking. Another Rai lives in Kathmandu and works in a carpet factory. One younger brother of a male household head serves as a policeman in India but doesn't send money to the family. None of the Rais have joined the British Gurkhas, nor the Indian or Nepalese armies.

Several clans are represented in the village and they establish who is and is not a potential marriage partner and assign funeral and other obligations. Proscription of marriage occurs within one's own clan and up to seven generations back with members of other clans (McDougal 1979). A collection of headmen represent the senior members of local clan segments. Houses are usually grouped according to the people's respective clans.

The Rais practice an indigenous religion minimally influenced by Hinduism. There are three *dhāmīs* (shamans) and eight *pūjāris* (priests) who are the religious practitioners for the village. Both *dhāmīs* and *pūjāris* perform rites to ensure the health for all household members and to treat illnesses. The *dhāmīs*, however, are responsible for conducting the major rites for the village.

Table 6 shows the census comparison for the entire research study area which includes the villages of Gongtala, Dobatak, Saisima and Yangden. The percentage age composition was taken from Tables 1-4.

Table 6

Census comparisons of research study area

	Age composition (%)				
	< 5	6-14	15-45	46-64	65 & >
Gongtala	26	23	45	6	-
Dobatak	9	17	54	14	6
Saisima	16	40	36	8	-
Yangden	15	29	43	7	6

There is a lower percentage of children in Dobatak because two married sons have no children and there are four nuns. Gongtala has the highest percentage of young children of all the villages. It is important to note that the sample size is small and the village populations are of varying size. The gender division of labor will be discussed in Chapter 4.

CHAPTER 4

Household Food Security

Forest owners have long used forests and forest plants, buried their ancestors in these forests and named the hills, rivers, and valleys where forests are found. Most forest-dwelling people are farmers. They grow cereals, for example, maize or root crops, but regardless of their main crop, they depend on forests for material goods, food and the definition of identity.   *(Alcorn 1996:235)*

This definition of forest owner explicitly depicts the marginalized Himalayan farmer, particularly the Rai, in the Apsuwa Valley. Alcorn (1996) further describes the situation of these forest owners, in that they live primarily in 'marginal' or 'peripheral' areas containing remnant forests, which cover about 20-70 percent of the community land.

The historic Rai ties to the *jaṅgal* are well-documented. "The Kirat of the north [the Rais] and the Limbu of the south were known to the ancients by the name of Kirāta, on account of their living by hunting and carrying on trade with the natives of the plains in musk, yak-tails, shellac, cardamoms, etc., from the earliest Hindu Periods" (Das 1970[1902:3n]). In the ancient Sanskrit literature, the Rais were known as remarkable warriors who fought with their bows and arrows (Sharma et al 1991). Nowadays the terms, *śikāri*, hunter, and *jaṅgali* people are often used by the Sherpas and high caste Brāhmans to describe the Rais. Other Tibeto-Burman groups are also dependent on the *jaṅgal*. As Kawakita (1984:12) noted, the Magars, a Tibeto-Burman group, are very knowledgeable about the ecological functions of the forest, "they feel an affection, exhibit a sort of pious attitude towards the *jaṅgal*, and enjoy their unrestrained life in the forest and high-altitude pastures."

Poverty Indicators

Poverty is increasing in Nepal due to unequal access to resources and land tenure policies (Amatya 1988). Nearly half of the total population of Nepal, about nine million people, are estimated to be below the poverty line (NPC 1991:2). A marginal farmer in Nepal is considered to own less than 1.02 and 0.20 hectares of land in the Tarai and middle mountain areas, respectively (Karan and Ishii 1994:70). With such small holdings, the marginal farmers are as vulnerable as the landless people because they can not meet their subsistence needs even with improved technology and increased cropping intensity (CBS 1985). A high level of indebtedness exists among the landless and near-landless households (UMN 1987). In a rapid appraisal of the Okhaldunga Community Health Programme in 1986, 62 percent of the households (178) had at least one debt, and interest rates ranged from 0-60 percent (UMN 1987). The average per capita agricultural landholding was reported to be 0.12 hectares in the hills (CBS 1987), which explains why emigration from the middle mountains has continued. Food deficits are growing in the middle mountain areas and malnutrition is increasing among low income groups in the region (Karan and Ishii 1994:66).

Throughout the Sankhuwasabha District, Eastern Nepal, the marginalized Himalayan farmer lives in dire poverty. Many outsiders perceive farmers as uniformly "poor," however, intra-village wealth and status differences exist (Barlett 1980). Studies in Eastern Nepal, between 1977-1981, revealed that both wasting and stunting in children from one to three years old were most prevalent among farmers who cultivated small areas of land. Particularly at risk were children from households cultivating less than half a hectare (Nabarro 1982). Stunted children[21] have a higher risk of mortality compared to children whose height is close to the reference value (Nabarro et al 1987).

Amount of Cultivated Land

The Apsuwa and Waling sub-catchment was rated seven out of twenty-one sub-catchments as having the least amount of agricultural land within the Arun Basin (Shrestha 1989:55).

Table 7

Land use for Apsuwa and Waling sub-catchment compared with the entire Arun Basin

	Apsuwa & Waling Sub-Catchment	Arun Basin
	17,548 ha	502,834 ha
Forest land*	53.83 %	50.48 %
Agricultural land	4.11 %	25.31 %
Grassland	20.41 %	9.95 %
Rocks and glacier	21.63 %	14.25 %
* including shrubland		

21 Stunted children have been defined as those children with a height less than 85 percent of the reference height for their age (Nabarro et al 1987:50).

The ratio between agricultural and forested lands is considered a useful criteria to evaluate the general condition of a hill region. Using this criterion, one hectare of cultivated land, requires at least 3, 3.5, and 6 hectares of forest land to support it according to Shepherd (1985), Wyatt-Smith (1982) and Applegate and Gilmour (1987), respectively. The Arun Basin as a whole has a ratio of 1:2 (agricultural : forest land) while the Apsuwa and Waling sub-catchment has a ratio of 1:13. This indicates that although the amount of agricultural land in the Apsuwa and Waling sub-catchment is quite small (4.1 percent), the forest land is sufficient to supply fodder, firewood, timber and other forest resources for the population. In the Apsuwa Valley, a high population ratio to cultivable land, or an unequal distribution of land within the communities or between communities, may be the underlying factor accounting for the existing poverty, but the presence of the *jaṅgal* helps to compensate the poor.

Formal Education

In the 1995-96 Nepal Living Standards Survey, literacy rates were 37.8 percent for children six years and older, 52.2 percent for males and 24.4 percent for females (CBS:1996:56). In remote rural areas such as my research area, literacy rates are even lower.

There is a primary school, Grades 1-5, located in Yangden. More than twice as many boys as girls attend school (see Table 8). A high percentage of children drop out after Grade 1. Although the annual enrollment is about fifty students, on a daily basis, few children attend school. On several occasions, I saw only three to six students at the school. Harsh weather conditions and household labor demands usually lower the children's attendance. The eldest sons and daughters of a household rarely attend school because they are needed for herding, farm labor, younger sibling childcare and other domestic chores. Children from Gongtala attend school in Yangden while children from Dobatak and Saisima usually go to school in Tamkhu (Grades 1-10) and stay with kin relations nearby.

Table 8

1992 school enrolment in Yangden by gender and class level

Class	Males	Females	Total
1	20	11	31
2	5	2	7
3	4	-	4
4	4	1	5
5	1	-	1
Total	34	14	48

A Rai girl carrying a baby in a *kokro*.

Household Food Security

Household food security is defined as "a household's access to an adequate year round supply of food" (Falconer and Arnold 1991:1). According to Velarde (1990:13), three conditions are necessary for household food security:

1) availability in quantity and quality of food supplies which are culturally acceptable;

2) accessibility of households to these supplies and stability of food supplies to ensure availability; and

3) access to them throughout the year.

The seasonal occurrence of inadequate food supplies, referred to as chronic food insecurity, is a permanent recurrent feature of traditional systems to which the population must adapt if they are to survive. Thus food deficiency (hunger) and poor nutritional status are rooted in poverty (FAO 1992).

Sankhuwasabha District is a food deficit area because current agricultural production cannot meet the minimum calorie food needs of the rural farmers (Cassels et al 1987). For high-altitude dwellers, Tibetan groups and Sherpas, livestock ownership is an important means to supply food products and as a source of barter by those farmers who have an insufficiency of grain crops.

Household food security is an ever present problem in this research area because even though there are no landless households, land is insufficient to provide for household food security. Also, the number and type of livestock owned by farmers cannot remedy the deficiency in agricultural food supplies. *Anikāl*, insufficient food supplies or a hunger period, then occurs as shown in Table 9.

Table 9

Length of *anikāl* experienced by households sampled in
Gongtala, Dobatak, Saisima and Yangden

	Number of Months of *Anikāl*					
	0	**1-2**	**3-4**	**5-6**	**7-8**	**9 & >**
Gongtala (10)*	2	2	6	-	-	-
Dobatak (5)	1	1	3	-	-	-
Saisima (4)	-	3	1	-	-	-
Yangden (32)	-	2	5	14	9	2
Total:	3	8	15	14	9	2

* Number in parentheses is the number of households in the village sample.

For the Yangden Rais, for example, *anikāl* averages six months per household (range of 2-12). Livestock and land-ownership accounts for the inter-household variation. In the Sherpa villages, *anikāl* averages 2.5 months (range of 0-5). The main months of food scarcity occur just before the next harvest during the pre-monsoon months (mid-February to mid-May). The three Sherpa households which reported no *anikāl* have a regular income from seasonal trekking. One Sherpa household which reported the longest *anikāl* of five months (among Sherpas) is considered the wealthiest family in Gongtala and was not included in the overall average. This household has the largest number of high-altitude livestock and is the main money lender in the area.

Land Ownership

Land is an important symbol of wealth, social status and power within marginalized communities in Nepal. The ownership of land is directly correlated with the standard of living, that is, the larger the landholding, the higher the standard of living. Questionnaire data is not infallible and landholding information which is often obtained to determine a household's economic status can pose problems. People may have been suspicious about motives for questions regarding land ownership and deliberately given misinformation, especially in areas where Cadastral Surveys had not been completed. In addition to this, people often do not know the actual area of land holdings (Fisher 1987).

In order to avoid suspicion of my role as a researcher, I used the amount of maize seed the farmers sow on their fields to roughly determine landholdings. This measure is based on the Bijan system which was used in hill districts in Nepal between 1933-1948. The standard unit used was a cubic measure of maize seeds, usually the *mānā* (about one pound). According to this system, land revenue taxes on *pākho* (unirrigated highland fields where only maize and millet dry crops can be grown) were assessed on the quantity of seeds expected to be needed for sowing. A major focus of this system is that it has no direct correlation with land area because acreages are not known and titles and boundaries are not always clear (Regmi 1978).

Households in my research area were asked how much maize seed they use for sowing. The data was not used as a strict criterion of land ownership because even this method aroused suspicions among a few farmers. Another reason is that Rai and Sherpa farmers may sow other grains on their land, in addition to maize. One wealthy Sherpa farmer stated his family used only five *pāthi*[22] (forty pounds) of maize seed although his landholdings are quite large and he often employs wage laborers to work his fields. In Gongtala, the amount of maize seed reported by households ranged from two to eight *pāthi* (16-64 pounds). There were a few correlations that could be made, for example, families which reported using only two and three *pāthi* of maize seed are poor, often borrowing money or working as wage laborers. A Sherpa family which uses eight *pāthi* (64 pounds) of seed is considered one of the wealthiest families in the village.

In Yangden, the amount of maize seed sowed varied, from one *mānā* to ten *pāthi* (one to eighty pounds) based on 28 households. The extended family that sowed ten *pāthi* of maize seed consists of three brothers living with their parents so the land holdings are quite large. One elderly woman living alone reported she uses only one *mānā* of maize seed and experiences eleven months of *anikāl*. The incomplete extended household, consisting of a grandfather and a granddaughter, reported they use four *mānā* of seed and experienced seven months of *anikāl*. Six

---

22 One *pāthi* is equivalent to eight *mānā*.

households in Yangden said that at least one family member had gone to Sikkim to farm because there was not "enough food" for the family.

Subsistence Agriculture

I used the farming system approach to look at all the economic activities employed by the entire rural community. The diversified farm enterprises used by the Rai and Sherpa farmers are characteristic of a regenerative agriculture system.

Rai and Sherpa Cropping Systems

The Rais depend on crops for the majority of their nutritive needs. The lower elevation and milder climate prevailing in Yangden allows the Rai farmers to grow a diversity of crops, however, the main ones are maize and millet (*kodo, Eleusine corasana*). Crops unique to Yangden, but not the higher elevation Sherpa villages are: buckwheat, *pustakari (Phul tarul*, a root crop), millet, soybeans, chili, Japanese-introduced *Pī ālu (Alocasia indicum)*, cucumber, mustard (*rape, Brassica spp.*), *philiṅgo (Guizotia abyssinica*, a type of black oil seed plant), sugarcane and *lungkupa (Amaranthus caudatus* L.). *Lungkupa* is used by Rais for making *jāḍ* (fermented beer). Dobatak is also able to grow soybeans and millet because it is located at a lower elevation. Black gram (*mās, Phaseolus mungo*) is also grown.

In Gongtala, high-altitude barley, naked barley, wheat, maize, potatoes, and radishes are cultivated. At higher altitudes, marginal lands with low fertility can support potatoes which are planted on land previously plowed by oxen or hoed by hand at the edge of terraces. The presence of five types of potatoes cultivated in Gongtala is evidence of the existing genetic diversity. These types are:

1) *riki sirbu* (S), *pahelo alu*, yellow potato;
2) *riki murbu* (S), *rato alu*, red potato;
3) *kumbe riki* (S), *khumbule alu;*
4) *riki karanchi* (S)
5) *riki karmu* (S), *seto alu*, white potato.

The potato is not native to the Himalaya and Furer-Haimendorf (1964) postulates that potatoes may have been first cultivated in the Everest region during the mid-19th Century.

As a non-cereal food crop, the yield of potatoes is two to three times greater than cereals. In addition to this, potatoes provide more energy and protein per hectare than cereal crops (Ahmed 1994). A major drawback for the farmer with few family laborers is that potatoes have the greatest labor requirement (excluding processing) of all the crops (Ashby and Pachico 1987). Potatoes are the main staple for the Sherpas during the summer months until maize ripens in the autumn. A Sherpa woman said, "I am like a cow because I eat potatoes all day long," referring to the monotonous grass diet of the cows. When potatoes are the main component of the diet, the labor demands of women are minimal compared to the heavy work required to process grains. The role of women and food processing techniques will be discussed in Chapter 7.

The economic relationship between Rais and Sherpas is mutualistic because their altitudinal stratified crops are bartered. One good is exchanged for another according to what the host village views as a fair rate (Humphrey 1992). Sherpas barter garlic and potatoes for millet, unhusked rice, maize and chili with neighboring Rais from the villages of Yangden, Yaphu, Chhoyeng, Waleng, Mengtewa, Tamkhu and Benchong. Humphrey (1992) asserts that bartering creates links and establishes friendly relations which cross ethnic and political boundaries.

All *pākho* or *bāri* fields are unirrigated, rain-fed terraces, constructed on steep slopes. The terraces are built with a slight inclination outwards to prevent water-logging caused by summer rainfall. The pre-monsoon precipitation which occurs during April and May is an important precursor for planting maize, millet and potatoes (Shrestha 1989).

Inter-cropping is an indigenous technique which ensures soil fertility and acts as a technique of pest control. Crop losses caused by insects can be severe. I knew a Rai farmer whose entire maize crop was destroyed by insects. Field fallowing is characteristic of agricultural lands on the hillslopes. In Yangden, maize, millet and beans are inter-cropped for one season; then the land is allowed to remain fallow for a certain period of the year. The nitrogen content in the soya plants is high and when inter-cropped with maize, the maize yield is significantly increased. Soybeans yield more usable protein per acre than other legumes, cereals or other cultivated plants (Simoons 1991) and provide more protein per hectare than potatoes (Ahmed 1994). Soybeans contribute four to eighteen times more usable protein per acre than an equivalent amount produced by milk, eggs or meat (Simoons 1991). For the Yangden Rais, soybeans are a low cost source of protein and an important source of oil.

In Gongtala, other combinations of crops are: beans, *Pī ālu* (*Alocasia indicum*), maize, garlic and onions; *Pī ālu* and potatoes; and naked barley, wheat and barley.

Double cropping is an indigenous technique to increase production. The primary summer crops are maize and millet, while the principal winter crops are barley, naked barley and winter wheat. Potatoes, *Pī ālu* and soybeans are planted in the early spring when the ground is not cold.

Fertilizer

A symbiotic relationship exists between the livestock and farming system. Plant wastes are recycled after harvesting millet because the stalks are a source of fodder. A gate is set up around the fields to keep the animals in the area before the seeds are broadcast. The animal dung and urine may be the main, if not the only source of fertilizer. Local fertilizer is a mixture of about one part of animal manure to three parts of leaf compost (*jangal* litter, grass and tree leaves) which are firstly degraded in the cow shelter and later applied to the fields. The amount of fertilizer needed varies from 50-100 *doko* (large, loosely-woven open baskets) depending on the size of the fields and type of crops grown. Garlic and potatoes require a lot of fertilizer and are grown by Sherpas who own a greater number of livestock than the Rais.

The ownership of only a limited number of livestock has nutritional implications. Farmers often complained that due to a lack of fertilizer, their crops did not grow well because the soil was not fertile. The need for manure creates internal trade because the use of oxen for plowing or manure is paid in grains or exchanged for labor. One Sherpa bartered fertilizer for maize stalks from a Rai who had no animals. A family which has a surplus of fertilizer will often give it to a family which has no livestock. Once the field on which it is used is harvested, half of the crop will go to the family which gave the animal dung. Oxen are loaned to plow the fields in exchange for either farm labor or the equivalent in grain.

Livestock Ownership

Another indicator of poverty or wealth is the number and type of livestock owned. Sherpa households own more livestock (from 5-59 animals) on average than the Rais (from 1-29 animals). The Rais have mainly sheep, goats, cows, oxen and pigs while the Sherpas also own *caūrī* (yak-cow hybrids) and other high-altitude animals. Owning too few animals can increase food insecurity for the farmer and put him or her into debt because fertilizer must be purchased. Usually households experiencing *anikāl* of eight months or greater duration own only one chicken, pig or goat.

The role of livestock varies according to the animal. Oxen supply draught power and manure. If a farmer borrows oxen to plow his field, he then donates his labor to plow that farmer's field. Cows produce milk products, and if they die "naturally," are consumed. Sheep produce wool, approximately one kilo of wool per animal. The Rai farmers said if the sheep are kept at the *lekh* or the *jangal*, their wool is thicker, and more is harvested than if the sheep are kept in Yangden. They are a source of income. One Rai man carried a large basket of food up to the *lekh* (high pasture), and in exchange for his labor, he was given five kilograms of sheep's wool, enough to make one woolen blanket (*rāri*). This wool would be the equivalent of five to six sheep kept at the *lekh*, or eight to nine sheep kept in Yangden. Goats are a source of income, sold for sacrificial offerings, funerals, weddings, and religious festivals, such as *Dasaī*, the festival of the goddess Durga. During the month of *Dasaī*, Rais came and took nine goats worth Rs 11,050 from Gongtala. Roosters are mainly used as sacrificial offerings to deities. Eggs are a source of income, used in ritual offerings and added by Sherpas to alcoholic beverages for prestigious guests. Dogs are used to herd the animals and are pets. When one Sherpa *lama's* dog died, he burned five butter lamps at the *gönpa* in his memory.

Pigs are not kept in Saisima. When the head *lama* at the *gönpa* in Saisima was asked, "Why aren't there any pigs in Saisima?," he laughed and answered, "Because there are *banel* (wild pigs) in the *jangal.*" Another Sherpa's response was, "Pigs are not kept in Saisima because you can't kill them and they would be difficult to move to another village through the *jangal*." Pigs produce income: 2.5 kilograms of pork sells for Rs 100. They are sold for weddings, *Dasaī* and *Losar* (T, *Lo.gsar*, the Tibetan New Year celebration). Pig bristles are bought by Limbus who come from Tehrathum, Hile and Dharan, the major market centers of Eastern Nepal. One kilogram of bristles sells for Rs 400. The bristles are exported to India where they are used for brushes and boat wheels. Pigs are fed grain husks, the dregs from the brewing of alcoholic beverages, kitchen refuse, *pustakar* (*Phul tarul*) and *Pī ālu* leaves (*Alocasia indicum*). Pigs therefore do not compete for *jangal* resources. Owning high-altitude livestock is considered an indicator of wealth because of the income earned from selling the animals and their by-products. They produce butter, an expensive food commodity: 2.5 kg of butter sells for Rs 200 or is bartered for ten *pāthi* (40 kg) of maize. The high-altitude animals produce a greater amount of milk and wool than cows. The nutritional considerations of this will be discussed in Chapter 7.

An additional nutritional consequence resulting from inadequate numbers of livestock is that fewer milk products are produced, which in turn lessens the amount of protein and animal fat available in the diet. Milk production fluctuates seasonally with fodder availability and is at its lowest during the winter months when fodder is scarce. Women, who are mainly responsible for fodder collection, are burdened with the increased work load of stall-feeding the animals. The decrease in milk production takes an additional toll on women who are dependent on the consumption of milk products (cheese, buttermilk and yoghurt) to help meet their own nutritional requirements.

Fodder

The villagers are very dependent on the *jangal* for fodder. According to Wyatt-Smith (1982), fodder needs are three to five times greater than fuel and timber needs combined. Private fodder trees and crop residue supplement *jangal* sources of fodder. During the three winter months when the ground is covered with snow, the animals are not able to free range. Corn stalks, from their own fields or bartered from the Rais, and leftover alcoholic remains are given to the animals by the Sherpas. Also, livestock problems – for example, leeches biting the animal's eyes and blinding them or cuts on animal's hoofs – prevents the animals from foraging and the farmers must bring them fodder.

The amount of time for fodder collection varies depending on the household's proximity to the *jangal*. Households reported it takes from one hour in Saisima to as much as six hours in Yangden to collect a *doko* of fodder. Approximately 35-50 *dokos* of fodder per week are needed, depending on the number of livestock owned.

Transhumance

Transhumance (pastoral movement) of sheep is practiced by both Rais and Sherpas. High-altitude bovids (yaks and their hybrid offspring) are solely herded by Sherpas. Several ecological zones are exploited during transhumance at altitudes beyond the upper limit of cropping (3,000 meters) where the forest and pasture areas begin. Pastoral transhumance involves a complex social organization, including individual and group decision-making and land tenure practices in order to assure that the animals have access to grass and water. Land use practices are based on the support of particular kin roles and other features of social structure such as age set systems (Furer-Haimendorf 1964).

Climatic conditions determine the transhumance strategy because the high-altitude livestock are moved to high and low pastures ranging from 2.700 to 4,300 meters over an annual cycle. The livestock are taken to the *lekh* (a high ridge that is not perpetually snow-covered) located in the uppermost Apsuwa Valley during the monsoon season (mid-June to September). They are taken to lower pastures in the winter months (October-May). During winter for about ten days, the cows, goats and sheep, are moved to bamboo sheds (*goⴕhs*) near the households where a fire is kept burning all night by a young herder who sleeps with the animals. The livestock are then moved to a lower pasture below Gongtala for one month where the weather is not as cold. When the animals are moved throughout the village, they must be watched closely because they tear off the unharvested grains. The Rais have woven muzzles made out of *mālingo* (*Arundinaria aristatla*) to prevent the livestock from eating the crops.

I once observed a Sherpa friend suddenly set a grassy area on fire while she was tending cows at about 2,700 m. The seasonal burning of grass improves plant reproduction by increasing herbaceous production and replacing older and less edible grasses with younger and more edible ones (Dove 1984; Sauer 1969). It also prevents the invasion of insect pests (Burton et al 1986).

This acts as a safeguard for unpredictable and static environments. Also, the sale of milk products and livestock supplements the subsistence base. However, this agropastoral strategy is possible only if labor is available. An elderly Sherpa expressed concern that when his young daughters get married, he will have to sell his high-altitude animals because there will be no one to look after them. A trend is occurring in this area, as well as other areas throughout Nepal, where more children are attending school and young people are leaving rural areas to seek "attractive and glamorous" trekking or wage labor jobs in Kathmandu. They don't want to be assigned the low status and lonely job of herding in remote areas for long periods.

The trading connection is still a viable and economically-profitable endeavor for Sherpas owning high-altitude animals in my research area. Pahakhola is a Tibetan trading village about a four-day walk northeast of Gongtala, located in a restricted area on a direct route to Tibet. For example, two Tibetan men came from Pahakhola selling woolen back aprons worn by Sherpa women for Rs 500 each. They bought two adult high-altitude animals and three calves. The trans-national trade which has been occurring for centuries with Tibet usually involves high-altitude livestock and firewood from Nepal bartered for Tibetan salt; kutki (*Picrorhiza scrophulariiflora*), a medicinal plant; tea and woolen cloth (Humphrey 1980).

Production Activities: Gender Patterns

In the research area, labor division is not strictly defined by gender nor social groups because all family members, men, women and children, contribute to a family's subsistence. The domestic and subsistence activities overlap and these tasks depend on the ratio of dependent to non-dependent workers, the health status of laborers and the pool of village residing household members. Anderson and Mitchell (1984) remarked how every child above the age of five is expected to do some chores, and even four- to five-year-old children help with household tasks, such as chasing chickens away from eating the drying grain. I saw children six years of age watch goats and cows. A time allocation study conducted by Nag et al (1978) found that Nepalese children, six to eight years old, spent an average of 2.2 hours on animal care and 1.7 hours on child care. More demands are placed on girls, as evidenced by the school attendance records for Yangden shown in Table 8. The higher education opportunities for boys is apparent throughout Nepal (Anderson and Mitchell 1984). The high level of children participating in economic activities is characteristic of traditional economies.

This research area does not fit the normal picture of dependency ratio which assumes that able-bodied workers are those between the ages of 15 and 65 years of age. Rather than following the usual category in showing the household composition, I used a framework which used a category of children, 6-14 years of age, because they engage in some economic activities (see Chapter 3, Tables 1-4), and are only age. Rather than following the usual category in showing the household composition, I used a framework which used a category of children, 6-14 years of age, because they engage in some economic activities (see Chapter 3, Tables 1-4), and are only considered partially dependent. For comparative purposes in the household economy, nuns work part of the year because it is not economically feasible to have full-time monks and nuns. Also, usually every year the *lamas* from Gongtala and Dobatak take off one month from farm work and herding activities to participate in the teachings at Dechen Chöling *Gönpa* in Saisima. This contrasts to the situation in Tibet where neither monks nor nuns work. Seasonal migration, another factor, places a heavy burden on women. At this point, the 6-14 year age group becomes particularly important to carry out many of the additional chores. Elderly people who live alone also work and only those who are frail or quite ill are excluded from this ratio.

Collecting 'wild' edibles and water and tending the fire are activities performed by men, women and children. Women perform the majority of the agricultural tasks, such as hoeing, weeding, harvesting the crops, selecting seeds for cultivation, placing manure on the fields and winnowing, husking and grinding the grain into flour. Men, however, are often seen assisting in these tasks. These activities will be discussed in detail in Chapter 7. Milking animals and the production of other animal products, cutting loads of grass and fodder for the livestock, carrying food to the fields for communal work parties, cooking, cleaning and washing clothes, are predominately female tasks. Herding animals is usually done by unmarried women, especially the youngest daughter-in-law, and teenage boys. The choice of herders depends on the family structure and size, along with which family member is available.

Plowing, chopping wood, collecting *māliṅgo* (*Arundinaria aristatla*), trekking, treading sheep's wool blankets to compact the weaving, buying trips for food supplies and salt, the construction and maintenance of terraces and the construction of houses are predominately male tasks. Sherpa men also travel to neighboring villages to collect grain or cash on outstanding loans.

Household Food Security Issues

Issues of household food security needs to be kept in perspective. The use of ingestible *jaṅgal* resources must be related to the acquisition and transformation of other components of the food chain. Therefore a seasonal calendar (Figure 3) was produced assessing the household food security situation. The data were gathered from the local people regarding their yearly planting and harvesting schedules, livestock management and those other annual cycles which have a major impact on their lives.

Figure 3: Seasonal calendar for Gongtala, Dobatak, Saisima and Yangden.

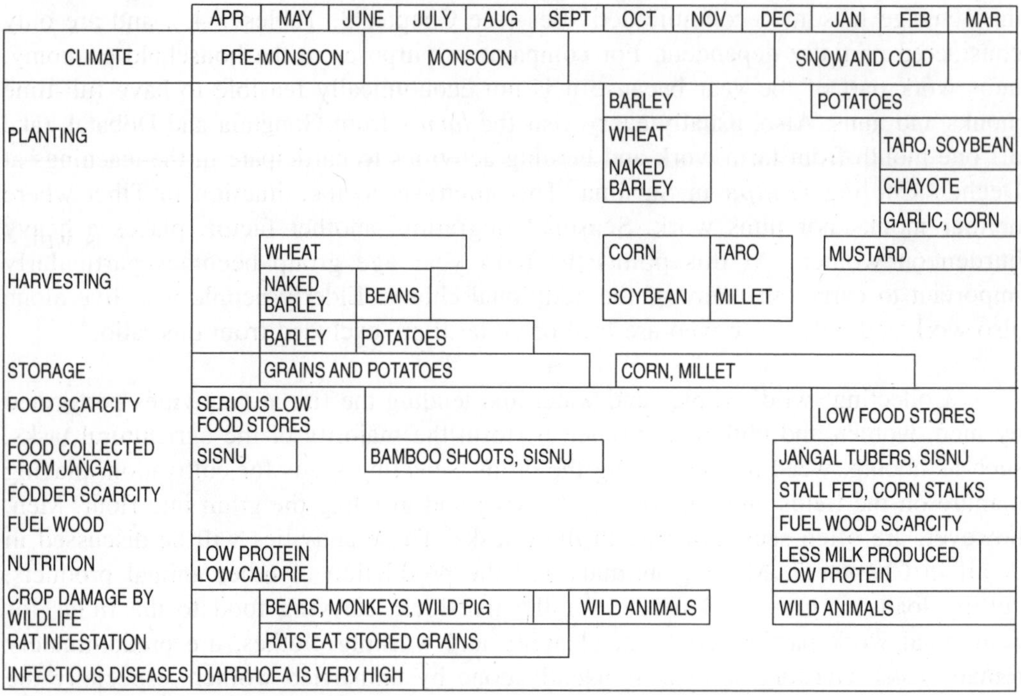

	APR	MAY	JUNE	JULY	AUG	SEPT	OCT	NOV	DEC	JAN	FEB	MAR
CLIMATE	PRE-MONSOON			MONSOON						SNOW AND COLD		
PLANTING							BARLEY / WHEAT / NAKED BARLEY			POTATOES	TARO, SOYBEAN / CHAYOTE / GARLIC, CORN	
HARVESTING		WHEAT / NAKED BARLEY / BARLEY	BEANS / POTATOES				CORN / SOYBEAN	TARO / MILLET		MUSTARD		
STORAGE		GRAINS AND POTATOES					CORN, MILLET					
FOOD SCARCITY	SERIOUS LOW FOOD STORES									LOW FOOD STORES		
FOOD COLLECTED FROM JANGAL	SISNU		BAMBOO SHOOTS, SISNU							JANGAL TUBERS, SISNU		
FODDER SCARCITY										STALL FED, CORN STALKS		
FUEL WOOD										FUEL WOOD SCARCITY		
NUTRITION	LOW PROTEIN LOW CALORIE									LESS MILK PRODUCED LOW PROTEIN		
CROP DAMAGE BY WILDLIFE		BEARS, MONKEYS, WILD PIG				WILD ANIMALS				WILD ANIMALS		
RAT INFESTATION		RATS EAT STORED GRAINS										
INFECTIOUS DISEASES	DIARRHOEA IS VERY HIGH											

Note: Respiratory illnesses also occur, particularly during the winter months.

The calendar illustrates the major problems which villagers face and how they cope with their situation. For example, it shows that *anikāl* occurs during the pre-monsoon period (April to mid-June) which coincides with the abundance of *jaṅgal* ingestibles such as *sisnu*, *jaṅgal* tubers, and edible bamboo shoots (referred to as *tusa*). There are three types of *sisnu*: *Girardinia diversifolia*, *Girardiana palmata* and *Urticia dioica* L. (see Chapter 5). One Rai farmer talked about their dire situation and the importance of the *jaṅgal* as a source of food.

> *Sometimes in Yangden we can't grow enough food. So we must go outside and buy food. There is no bridge at the river and during the monsoon season the river swells up, making it impossible for us to cross it. We must go to the jaṅgal to find sisnu. We feed our children the sisnu to eat with chakla.*[23]

The importance of *jaṅgal* ingestible plants and *jaṅgal* resources in alleviating the stresses experienced during *anikāl* will be discussed in Chapters 5, 6, 7, 8 and 9.

---

23 *Chakla* , a main staple eaten by the Rais, made of coarsely ground maize which is boiled.

Unpredictable Events That Aggravate *Anikāl*

Unpredictable events precipitate additional hardships during *anikāl*. Wildlife damage to crops, wildlife predation on domestic animals and food lost during storage worsens the *anikāl* period. Natural hazards aggravate the household food security situation. In the Arun Valley, landslides, torrential rains which frequently occur during the monsoon season and extremely cold weather create additional environmental hazards causing a loss of land, life, livestock and food supplies. The bridges in the area are unsafe and there are many stories of villagers slipping off the log bridges and drowning. Illness or death of household laborers are stresses which the farmer must also confront on a daily basis.

Wildlife Predation and Damage to Crops

Wildlife are a menace to the farmers because they often eat the ripened crop before it is harvested and prey on their domestic animals.

Table 10

Crop damage and livestock predation

Wild animals	Crops	Predator of
Chil (eagle)		chicken
Bāgh* (tiger)		dog, goat
Wild dog		cows, goats, sheep
Bear (sloth)	maize	
Wild boar	all crops	
House rats	all crops	
Rhesus monkey	maize, beans, millet	
Lokharke (Palm squirrel)	all crops	
Mirga (barking deer)	all crops	
Jungle cat		chicken

* Local people use the term *bāgh* in a general way to refer to any large wild cat.

The farmers use traditional ways of confronting the wild animals because they do not hunt nor kill them. Chickens are placed in bamboo baskets or brought into the house at night. After *pī ālu* (*Alocasia indicum*) are dug up, they are hidden in the ground, covered with brush and dirt and stones are placed on top. According to Rindos (1984), this is an adaptive strategy because many species of yams and sweet potatoes decay very rapidly once removed from the ground. Prior to the maize harvest, both women and men must stay in a bear shelter (*yarsa, [T, dpyar.sa)* and burn a fire all night in order to scare away bears and other wild animals. A bear can cause extensive damage. In one night, the farmers estimated that twenty *pāthi* of maize (about 80 kg) at Rs 25/ *p āthi* at the market price, can be destroyed, an equivalent of Rs 500. According to Krause's computations (1988), the annual consumption needs of one adult is two *mānā* per day (two pounds of grain per day) or (91 *pāthi* per year, 332 kg) and a child is considered one half an adult. Thus 20 *pāthi* (73 kg) of maize is equivalent to more than an additional three months of *anikāl* for one adult. Approximately 10-20 percent of the farmer's post-harvest staples stored in the loft are eaten by rodents (rats and squirrels).

A reciprocal exchange relationship among people who form labor groups referred to as *parma* is used in order to harvest crops quickly and prevent them from being eaten by wild animals. Beer is brought, along with snacks, such as potatoes and chili, for the laborers.

Livestock Loss

Winters can be extremely harsh. In Yangden during the winter of 1991-92, eleven sheep died because of the cold weather. Aged animals are kept in animal shelters (*go ʈhs*) near the households where a fire burns continuously throughout the night. Animals are moved lower than the villages where the temperature is slightly warmer. A social support network exists for households whose livestock have been killed. When a Sherpa's cow died because of cold weather in Gongtala, the meat was divided into equal portions and each household gave Rs 25. A total of Rs 300 was collected.

Health

The serious health ailments in remote farming communities debilitate families and cause lost labor and income. Fricke (1984) asserts, "A man or a woman's death can be an untimely interruption of the process of household building and radically transform the prospects of a household from a successful to a failing domestic economy." The timing of illness is crucial, especially if it occurs during the busiest season, when labor demands are at their peak. If a child becomes sick, the mother faces a dilemma: whether to care for her child or work in the fields.

The *dhāmī* (shaman), *lama* (Buddhist priest) and home remedies using medicinal plants, comprise the indigenous healthcare system. Modern medicine is expensive, and often not available. Infant and child morbidity and mortality seems to be high, based on the women's health interview which will be discussed in Chapter 7. According to Fricke's (1984) Tamang demography, one-fifth of the children will die in the first year of life while two-thirds will survive until at least the age of twenty. During my stay, I was often consulted by the villagers with a range of ailments. It was difficult to know which health problems the people had from their symptom descriptions.

Children suffer from skin conditions, sores and boils. I always got scabies during my stay and labelled it "the four-month itch." Broken bones are a common occurrence due to falls from trees while collecting fodder. I saw one Rai man who had broken his leg but still hobbled to work even though the bone needed time to heal. Cuts from *khukuri* (long knives) often become infected. Women suffer from sore and infected eyes because the traditional houses have no outlet for the smoke from the open fire. Many infants and children get burned from falling in the three-stone fire. During my stay, a ten-month-old baby fell in the fire when his eight-year-old sister fell asleep. He later died. Worms are endemic and everyone complains of stomach ailments. Diarrhea is rampant, babies roam everywhere and eat dirt. The people must work in the fields even when they have a fever because one day of lost work means less food brought to the family hearth.

Family Debt

A major problem among rural hill farmers is family debt. Family debt arises because marginalized farmers are not able to transfer their subsistence incomes (farms and farm products) into cash, nor can they easily obtain cash (Ives and Messerli 1989). Insufficient subsistence crops forces the farmers to use their cash reserves to buy staple grains. Money must be borrowed for buying clothes, salt, kerosene and school fees. Outside food commodities are bought for special occasions. For example, unhusked rice is not grown in the area and must be purchased for weddings or religious celebrations. Rice is a prestigious food item because it is expensive and not a normal staple. Rice sells for Rs 64 per *pāthi* (four kilos) in Khandbari, more than three times the village cost of maize (Rs 20 per *pāthi*). The villagers are further marginalized in the eyes of the dominant lowland group because they do not eat rice (see Seeland 1993).

Although salt is a major food commodity for the Rais, they have limited funds to purchase it. Salt is their main, and often only, condiment used in cooking. It is also needed for the livestock or they will get "thin." During the Nepali month of Mangsir (mid-November to mid-December), after the maize has been harvested, about thirteen to fourteen Yangden men travel to Hile for their annual salt buying trip. The entire round trip takes about thirteen days. One kg of salt sells for Rs 1.5 in Hile, and in Khandbari, it costs Rs 7, almost five times the price in Hile. The distance from Hile to Khandbari is four days which accounts for the higher price of salt in Khandbari to cover the porterage fees. In Tamkhu, the price of salt is Rs 14 per kg (a one-day walk from Gongtala).

Each man carries about 40-48 kg of salt. Due to limited funds, the Rais cook their own meals, carrying household food supplies and sleep outside. About 80-100 kg of salt lasts the entire year for a Rai household. Sherpas have a lot of livestock, so almost twice as much salt as the Rais must be purchased. Every two weeks, four kg of salt is needed for the cows and sheep. Salt is added to the Sherpa's Tibetan tea.

More serious financial obligations are incurred in animal sacrifices for prolonged illnesses, and for social events such as funerary expenses and the increased cost of marriage. Death rituals can be a major factor contributing to poverty. The complexity of the ritual and the number of people involved in the ceremony, is usually dependent on the web of social relationships and age of the deceased (Wahlquist 1981). One Rai man recalled the reason underlying his family's impoverishment. His father was the former *Pradhān Panchāyat*[24] for the research area and had been chosen because he was the only educated person at the time of the elections. Therefore many people were expected to pay respects to his family. His son recounted the events surrounding his father's funeral.

Nine years ago (1983), I had to sell my land because I needed cash for my father's funeral. I sold land to a Sherpa, the largest land owner in Gongtala for Rs 1,000, equivalent to six pāthi of maize seed. This land is Pakala Ḍǎḍā. Now the land belongs to his three sons. My father was sick and I needed the money for a pūjā. A sheep, goat and chicken were sacrificed to feed our clan, kin relations and guests. A large amount of grains were needed to make jǎḍ. I had no money and therefore had to sell my land. This is the reason why I am now a poor farmer.

For the majority of households, selling land is considered to be the last resort when all other strategies have failed (Cassels et al 1987).

Wedding costs can be expensive because of bridewealth payments and marriages may be delayed, depending on the families' financial status and whether it is an arranged marriage or by choice. Although the marriage is held, all the formal rituals may not be completed for many years or not at all.

The scale and cost of the marriage celebrations and ceremony may be influenced more by the family's status and expectations of the community than by the resources available to the households concerned (Cassels et al 1987). One wealthy Sherpa traded two goats (a large one for Rs 1,200 and a smaller one for Rs 600) with a Newar man for two large copper pots from Dingla Bazaar. The pots were to be used at his daughter's wedding for *chang* (fermented beer) and rice.

The ritual for the first marriage is very costly and demands resources beyond a household's means. Indigenous to the Rai culture is *sachhep* (Yamphu Rai) which consists of contributions collected among the guests at the wedding celebrations in the bridegroom's house (Wahlquist 1981). At one Rai wedding I attended, Rs 300 was collected from the guests to help allay wedding expenses. However, all the money was given to the seven Damai musicians (who are tailors and considered as an untouchable caste) for their services.

---

24 *Pradhān Panchāyat* is a headman of a village.

One additional aspect of impoverishment occurs when land fragmentation results in shares too small to support the household. This can often lead to the break up and dispersion of the family group. The land reform program has not met with success to extend agricultural credit to farmers (Cassels et al 1997).

Coping Strategies to Combat *Anikāl*

Households usually cope with chronic food scarcity by employing a combination of strategies. The type of coping strategy is often determined by a number of factors including the length of *anikāl*, the resources owned by the household (land, livestock and gold), the current availability of off-farm labor, and the health status of family members. As Cassels et al (1987) point out, the choices of strategies to pursue are limited, and intra-household variation occurs in the order in which these strategies are employed. The main traditional practices which the farmers use to respond to their precarious food situation are: social support networks, the sale of agricultural produce and livestock, craftwork sales, wage labor, seasonal or permanent migration, loans, and the collection of 'wild' plants and the medicinal plant trade. The type of adaptive strategy used during *anikāl* gives insight into the differences between the two communities and reasons why it is longer for the Rais (averaging six months) compared to the Sherpas (two-and-a-half months). These coping strategies will be discussed in the following section.

Social Support Networks

Social relationships established with kin, and with villagers outside the kin network, provide economic and social support that is important to all rural households, especially the poor and women. During crisis periods, these established relationships are the first to be sought out for cash loans and wage labor. They also may provide reciprocal labor-sharing arrangements during the peak agricultural season. During periods of food scarcity, small amounts of foodstuffs are borrowed from these established social networks. Marriage alliances which extend reciprocity relationships with other households are a form of non-kin social relationship (Fricke 1984). Daughters are given to carefully chosen men to establish or bolster alliances with other families or villages (Kohn 1992). These marriage alliances change the labor force and composition of both the receiving and sending households and the ratio of dependents to non-dependents. Marriage ceremonies take place during the slack period after the harvest when there is little work and sufficient food stocks.

In addition to their links with consanguineal and affinal kinsmen, Sherpas and Rais establish ritual friendships very similar to kin ties, referred to as *mit* for men and *mitini* for women. These friendships close gaps in the social network, provide trust essential for trade transactions, and provide security and aid in emergencies. A *mit* commitment includes the partners' entire pool of relatives, ritual avoidance of the partners' spouses, and conducting funerary rites for the *mit's* death ritual (Humphrey 1992). These social relationships are maintained by the frequent exchange of small gifts of food. A man may have an average of three to six *mit* relationships chosen from clans with which he does not have any affinal relations or from ethnic groups in other villages.

I saw a few instances of these networks in my research area. A milking cow was given as a temporary loan to a household for a woman who had recently given birth. A Sherpa family loaned a hen to a Rai family which owned no livestock, boosting the protein in the Rai family's diet. These loans acted as an immediate source of food for the destitute borrowers. One poor Sherpa woman (with no food nor cash reserves) borrowed two pounds of salt on a promised agreement that she would return the same amount of salt when her husband returned from trekking.

Individual households which are confronted with food shortages may use a strategy of inviting close kin relatives and not the entire village to special ceremonies. March (1987) asserts this is contrary to Sherpa culture where social life revolves around hospitality and presses people's resources and sociability to such an extent that they may choose to avoid one another completely. One poor Sherpa woman gave birth and according to the Sherpa custom, after three days, the *lama* came to the household for the baby naming ceremony. The household only invited their own clan and the other villagers who had not been invited felt slighted. The household faced a dilemma because large amounts of grains are needed to make the alcoholic beverages and they were already confronting the problem of *anikāl*. The household chose to forgo the hospitality which is the norm of Sherpa society in order to reserve food supplies for subsistence needs.

A monthly Buddhist celebration for Guru Rinpoche (Padmasambhava) referred to as *Daśamī* or Tsheba cu (T, *tshes.ba bcu*) is observed on the tenth day of the Tibetan lunar calendar in all the Sherpa villages. The houses throughout the village rotate on a volunteer basis with ritual contributions. Rice, maize, buckwheat pancakes, *roti* (unleavened flat bread), passion fruit, bananas and potatoes are examples of foods given to all villagers who attend this celebration. Statues referred to as *torma* (T, *gtor.ma*) are deity representations made of barley and decorated with flowers made of butter. The *torma* are offered to the deities and later eaten by the villagers. Butter is also required for the 108 butter lamps which are lit for the celebration

Table 11

Market price of foods used for the *Daśamī* celebration

Foods	Amount of Food	Market Price	Equivalent Cost (Rs)
Naked barley	1.5 *pāthi*	Rs 25/*pāthi*	38
Rice	3 *pāthi*	Rs 64/*pāthi*	192
Maize	1 *pāthi*	Rs 25/*pāthi*	25
Potatoes	5 *pāthi*	Rs 15/*pāthi*	75
Wheat	1 *pāthi*	Rs 25/*pāthi*	25
Butter	1.3 kg	Rs 80/kg	105
Barley	1 *pāthi*	Rs 25/*pāthi*	25
Chang	11 *pāthi* mixed grain	Rs 25/*pāthi*	275
Total:			**760**

These ceremonial expenditures serve to affirm or enhance the villager's commitment to principles of social interaction among kinsmen and/or neighbors (Berry 1980). Every household in the village has an obligation to hold the *Daśamī* celebration which is rotated monthly by mutual agreement. However, for a poor household experiencing *anikāl*, Rs 760 can impose a major burden to prevailing scarce resources. DeGarine and Koppert (1990) have deplored "the lavish use" of grain resources to brew beer and distill alcohol, viewing it as food wastage because it replenishes the household grain supplies. The author's example is a rebuttal to Miracle's (1961) remark that the inefficient conversion in terms of calories in the transformation of staples into beer fights the monotony of the diet. Miracle views the drunken state as an efficient method of temporarily relieving social anxiety, especially in regions where hardships are common. Another factor to consider is that the thick *chang* served by the Sherpas is rich in the B-complex, especially thiamine and niacin, thereby enriching the local diet.

Alcohol-making is an income-generating activity for both Sherpa and Rai women. The hospitality of serving alcohol coincides with the harvesting of maize. The Rais serve *jāḍ* on special occasions, for example, *Dasaī*. During this time, my Rai friends invited me to their homes because they were able to extend hospitality with the *jāḍ* and *raksī* (distilled liquor).

Religion also plays an important role in social network relations. The Buddhist philosophy, based on the doctrine of *karma*, teaches compassion to all human beings which helps attain merit, essential for desired reincarnations (Humphrey 1992). A well-off Sherpa household frequently gave a bottle of buttermilk or curd cheese (S. *serkam*) to a sick pregnant woman in a poor Sherpa family for good *karma*. This act is considered a moral activity to earn merit. A Sherpa brother gave his favorite sister a baby goat referred to as *pewa*. His sister often takes care of his animals for him. In Sherpa society, women inherit, own and control important property (March 1987).

Sale of Agricultural and Livestock Products

For the Rais, sources of cash income are minimal. The KHARDEP Programme stated there is a need to search for alternative sources of income and stressed the importance of off-farm work opportunities (Cassel et al 1987). The Rais barter *philiṅgo* (*Guizotia abyssinica*), *pustakar* (root crop), buckwheat and soybeans for rice, maize and salt with Sherpas and Rais from neighboring villages. Alcoholic beverages are sold. One Rai man sold five liters of *raksī* in Gongtala for eight rupees per liter, earning a total of forty rupees. One Sherpa man went to Yangden to buy *raksī* from a Rai woman with whose family he had already established a relationship. Pigs and goats are sold, bartered and/or traded with the Sherpas and other neighboring Rais for religious celebrations. The Sherpas' main source of income is from selling agricultural and livestock produce and seasonal trekking.

In extreme situations, households sell livestock as an emergency measure in order to obtain large sums of cash. Prices range according to the type of animal, for example, an adult goat sells for about Rs 500, a sheep for Rs 700, and a mother and a lamb for Rs 1,500. High-altitude animals, owned by Sherpas, brings a higher price, a *caūrī* (yak-cow hybrid) sells for Rs 5,000 and a yak for Rs 7,000. Although animal sales supply needed cash for the household, the economic and nutritional value for family members is curtailed.

Craftwork Sales and Trade in *Chiraito*

Woven bamboo products are labor intensive and yield little income. The highest source of income for the Rais comes from the trade in *chiraito* (*Swertia chirata*), a medicinal plant, discussed in Chapter 8.

Wage Labor (Cash and Kind)

The wage labor potential depends on the number of potential household laborers, the health status and availability of casual off-farm labor. Krause (1988) pointed out that "the giving and taking of labor is the main exchange between clients and their patrons and this also provides a focus for the articulation of caste differences." In my research area, the Rais are the clients and the Sherpas are the patrons in this wage labor relationship. The Sherpas consider themselves higher than the Rais on the caste hierarchy while the Rais view themselves on the same level as the Sherpas.

Rai men often take care of sheep owned by Sherpas. The payment comprises food in kind (*anna*) and *chang*. After one year, a sheep is given as payment. When the Rais come to Gongtala, they are fed and stay with the Sherpa whose sheep they tend because of this existing social and labor relationship. This is a reciprocal relationship for the Sherpas who stay with the herder's family when they visit their village.

The Rais often porter for the Sherpas, carrying grain or salt. One Rai man earned Rs 900 for carrying sixty kg of salt from Hile for a Sherpa. The Rais, and poorer Sherpa household members, work for the larger landowners in Gongtala or Dobatak. The type of work is gender differentiated. Women are very dependent on these informal social networks but often work for very low wages, 15 rupees per day, compared to 25 rupees per day for male laborers. Women usually harvest grain crops, while the men cut trees or break stones. A poor Sherpa woman worked for a wealthy Sherpa household and was paid Rs 15 for weeding potatoes all day, while two Sherpa men cut *māliṅgo* (*Arundinaria aristatla*) and were paid Rs 25 each. Men's work is considered of greater importance as noted by food-in-kind payments, which may be one and a half times more than that received by female laborers. A poor Sherpa woman received one basket of naked barley for two days of agricultural labor. One day of plowing is equivalent to four kilograms of potato seed or Rs 25.

Recruiting day laborers against payment indicates a basic inequality in the relationship. A large landowner in Gongtala hired two Rai sisters for several days a week to harvest his maize. The women are the wives of the Rai man who sold his land to cover his father's funeral expenses. The ripened maize is planted on the land which he had sold. Their payment was one *pāthi* of maize (4 kg) an afternoon meal and *chang*.

One poor Sherpa farmer in Gongtala who had only one goat, no fertilizer and little land was allotted a piece of land and fertilizer by a wealthy Sherpa landowner. Once the wheat had ripened, half of the crop was then given to the wealthy landowner.

Long-Term Migration

Migration has historic dimensions rooted in poverty. The Rais moved to Yangden from Mahakulung because of *duhkha* (hardship) associated with the difficulties of feeding their families on the small plots of land available to them. Inadequate land is now re-occurring seven generations later (182 years) and infertile land is again the driving force for the migration. The major destination of migrants is Sikkim. Six Rai households reported one adult male working in Sikkim. As Russell (1992) noted, Northeast Indian states exerted more of a migratory 'pull' on the inhabitants of Eastern Nepal than Kathmandu, the capital. The types of employment in Sikkim varies from truck drivers to agricultural laborers. Recently Kathmandu has attracted young unmarried Rais. Two teenage girls ran away in the middle of the night to seek work in Kathmandu. Three Rai men work in the carpet factories in

Kathmandu, six days a week, from 5 a.m. to 11 p.m. They receive Rs 400 per month in wages, two meals a day and a room. They said the expenses incurred in Kathmandu makes it difficult to save money. In my research area, Rai males have not been recruited into the Indian, British Gurkha or Nepalese armies.

The labor potential of the household determines the migratory patterns and as Nabarro et al (1989) points out, outward migration can only occur in households where there are two able-bodied adults. Long distance or seasonal labor migration creates changes in the gender ratio, increasing the female-headed households and dependency ratio in the local communities and organization of labor, as well as the balance of resources available to males and females. Incomplete nuclear households (female-headed) rely on kin networks for gender-defined tasks, such as plowing. Intra-household decision-making strategies change and the women's position becomes even stronger in her husband's absence. I saw several of these Rai women attend predominately male village meetings and represent their household votes.

The economic consequences of migration may be positive or negative. Several Rai women noted that they did not receive remittances from their husbands because life was also difficult in Sikkim. Remittances are received by two Sherpa families who have children who own and work in a restaurant in Kathmandu.

Seasonal Migration

Since Nepal first opened its doors to foreign visitors in the 1950's, Sherpas have had an international reputation as mountaineer guides and high-altitude porters. Although Sherpas from the Khumbu area (Mount Everest) have been involved in tourism since the 1960's, the involvement of Sherpas from Gongtala, Dobatak and Saisima is a more recent phenomenon. The trend is changing because a greater number of Sherpa men are trekking and for more extended periods of time. The women and children are thus responsible for all the household subsistence activities while the men are away.

Last March and April (1992), eight Sherpa men from Gongtala (out of eleven households surveyed) had gone trekking. The income received is evidenced by the recent transformation in the village architecture, from older houses made of bamboo mats to newer houses of stone and wood. Income from trekking does not always return to the villages. Several women complained to me that the money earned by their spouses is often spent on liquor and gambling in Kathmandu. In addition to their wages, the amount of a tip given by foreigners can vary greatly. One Sherpa said that for a twenty-three day trek an old Canadian man gave him Rs 10,000. In another instance, an American woman gave a Sherpa man Rs 20,000 to build a house and Rs 8,000 to buy his wife a pair of gold earrings. Yangden Rais don't go trekking and so do not have this as a subsistence option. They say that they are not adapted to the extreme cold because they do not have warm clothing.

Loans (Food or Cash)

Loans are most often taken from local money lenders for consumption purposes by poor farmers. These local money lenders are considered wealthy because they have abundant agricultural land and livestock and cash available for loans. They form a locally-controlled type of assistance existing in the informal economy (Wahlquist 1981). The author asserts that these credit transactions rarely create new social relationships but rather augment or reinforce already existing ones. They are among the most important elements making up the social fabric of the society. The borrowers initially are those individuals with whom the lenders maintain moral ties (Wahlquist 1981). Beer and fermented alcohol are firstly extended as a hospitable offering in all cash negotiations. The Sherpas liked the strong Rai *raksī*, a characteristic which could aid in the transaction being approved. Food items such as chili, bananas, cucumbers and corn on the cob are also given along with the alcohol.

One main feature of the Kosi Hills Area Rural Development Programme has been the establishment of the Small Farmer Development Programme to enable groups of poor farmers with small landholdings to obtain access to services and cheap institutional credit (Nabarro et al 1987). This program has not yet reached my research area in the remote upper Apsuwa Valley.

"The traditional relationship between highly-ranked givers (or lenders) and lowly-ranked receivers (or borrowers), is described as 'part of the Maussian gift-exchange structure of reciprocity'"(Ives and Messerli 1989:159). Fisher (1990:73) asserts these traditional patron-client ties had some positive merits in terms of enabling the poor to get through bad times, but it is easy to romanticize this and reciprocity can very quickly slip into debt-bondage. A survey found that about 24 percent of the total credit needs met by formal institutional credit sources are not easily accessible to small and landless farmers. Thus farmers are forced to depend on local money lenders for more than 80 percent of credit requirements at comparatively higher interest rates (Amatya 1988).

Krause (1988) remarks that between close relatives, loans are said to be given without interest. This is not in agreement with my findings. One Sherpa man told me that when he had borrowed money from his uncle to build a restaurant in Kathmandu, he was charged 25 percent interest per annum, which is the usual rate in this area. The Rais borrow money from the Sherpas in Gongtala at this same rate during crisis periods when there are insufficient cash reserves to buy food. Usually the money cannot be paid back and frequent collection trips are then made by the Sherpas to the Rai households. For years the Sherpas may receive salt or grains. However, due to the annual 25 percent interest rate, the amount owing remains more or less the same despite the regular payments (Cassels et al 1987). These interest payments create a continuous drain on resources. The Sherpas after arriving at the Rai villages are fed and given alcoholic beverages, in addition to the grains in partial payment of the outstanding debt.

Social status leads to prestige. An elderly Sherpa is the main money lender for Gongtala and neighboring villages. He is wealthy because he has a lot of land, livestock and sells butter, milk and cheese. One Rai farmer came from Yaphu two years ago to borrow Rs 4,000 and now wants to borrow another Rs 8,000. The first loan is still outstanding. In February, two Chhoyeng Rai brothers carried 10 *pāthi* (40 kg) of maize to the Sherpa money lender which they sold. They were not paid for portering, however, the money lender's wife gave potato seeds and garlic, and fed them a meal and *chang*. The Rais wanted to borrow Rs 2,000 so they could buy land, and promised to pay back the loan by the Nepali month of Saun (mid-July to mid-August) six months later. Two Chhoyeng Rai men came to Gongtala to borrow Rs 4,000 and said they would pay it back with an equivalent amount in salt or grain. They gave a gift of smoked fish in a small container made of plaited bamboo. They told me, "we came to the Sherpa village because we felt money is available."

CHAPTER 5

Description of *Jaṅgal* Resources

The vegetation of the Arun Basin is diversified within each altitudinal zone (Karan and Ishii 1994; Shrestha 1989):

- The **subtropical** zone (1,000-2,000 m) contains mainly moist hill pine but also a few oak forests.
- The **temperate** zone (2,000-3,000 m) contains evergreen montane broadleaf species forests of oak and deciduous trees. Bamboo species are abundant.
- The **subalpine** zone (3,000-4,000 m) contains pine, fir and birch forests and rhododendron (more than one meter high). Birch trees grow at the uppermost limit of the timberline at about 4,250 m.
- The **wet alpine** zone (4,000-5,000 m) is essentially a zone of herbs and grasses, however, thickets of junipers (*Juniperus recurva*) and dwarf rhododendron bushes, also grow. At the uppermost level, are high-altitude mountains and permanent glaciers.

The northern and western-facing forested slopes of the Arun Basin contain great genetic diversity. Shrestha (1989) remarked that the Arun Basin's richness can be shown by the number of rare and endangered species, including more than 3,000 species of vascular plants, and about 25 species of rhododendron, 80 species of fodder trees and shrubs, 60 species of medicinal plants, and a number of 'wild' edible plants. Burbage's (1981) extensive study of the medicinal herb trade in the Koshi Zone, Eastern Nepal, revealed 32 species of medicinal plants.

The term *lekh* refers to a high ridge that is not perpetually snow-covered and just above the treeline at about 4,300 m. *Buki* (*Kobresia nepulensis*), a sedge; and *sunpati* (*Rhododendron anthropogon*) are found at about 5,500 m at the *lekh*, which corresponds with the alpine zone. Vast meadows contain colorful, but unpalatable flowers, *Primula* spp., *Potentilla* spp., *Ranunculus* spp. and other herbs. Rajbhandari (1991) and Shrestha (1989) attribute the heavy grazing by sheep and yak to their abundance. *Jaṅgal* ingestibles are also gathered.

Table 12

Relationship between bioclimatic zones, altitude and *jangal* resources

Bioclimatic Zones	Altitude (meters)	*Jangal* resources
Alpine	4,000-5,000	mushrooms, 'wild' greens, incense plants, medicinal plants
Subalpine	3,000-4,000	firewood, fodder, mushrooms, bamboo, medicinal plants
Temperate	2,000-3,000	*chiraito* (medicinal plant), 'wild' fruits, bamboo, mushrooms, 'wild' greens, *lokta* (paper plant), fodder, *allo* (a fiber plant)
Sub-tropical	1,000-2,000	'wild' tubers, 'wild' greens, nuts, fish, firewood, fodder

Within the altitudinal zones of the Arun Basin – sub-tropical, temperate, sub-alpine and alpine – Rais and Sherpas exploit a variety of *jangal* resources: medicinal plants, various bamboos, fiber plants, incense, 'wild' fruits, 'wild' green vegetables, tubers, mushrooms, 'wild' honey, fish, birds, hornets and other insects. (Table adapted from Martens (1983), local knowledge and personal observations).

Disturbed Environments

In disturbed habitats altered by fire or swidden cultivation, *jangal* resources flourish. Bamboo growth is favored in the upper temperate zone (2,400-2,500 m) of swidden areas. Secondary successional forests on common property lands have been viewed by planners, policy makers and agronomists as "empty, unowned, useless and abandoned" (Hecht et al 1988:26). The non-crop vegetation on these lands are referred to as 'weeds,' 'brush,' or 'wasteland' (Alcorn 1995:2). The farmers do not perceive them as 'weeds' because they fulfil multiple uses for the household subsistence economy. *Chiraito* (*Swertia chirata*) a medicinal plant and important income source, is collected on swidden fields and rambling raspberries (*Rubus* sp.) grow in disturbed grounds around Gongtala. The *jangal* and *lekh* are also known for their fauna, which were identified by villagers in Table 13. Non-vertebrates in the Arun Basin include leeches, concentrated between 1,200 to 2,700 m elevation, and *bhusuna* (midges), both associated with the monsoon season.

Table 13

Fauna at the *jaṅgal* and *lekh*

Nepali	Sherpa	Common Name
Jhāral	*Rirabu*	Himalayan Tahr
Kasturi mrga	*La*	Musk Deer
Kālo bhalu		Himalayan Bear
Bāgh		Common Leopard, Tiger
Sokpu	Yeti	Abominable Snowman
Dānphe	*Dange*	Impeyan Himalayan Monal
Cilime	*Sermun*	Blood Pheasant
Lariwa	*Ti ling*	Wild Bird
Munāl	*Omung*	Crimson Horned Pheasant
Bwanso	*Mateng*	Wolf
Hunrunge,	*Taiyakpa*	Pika
Jarāyo		Sambhar
Kharāyo		Indian Hare
Jaṅgali bākhro		Wild Goat (Red)
Syaul		Jackal
Ghoral		Himalayan Chamois

Research Methods

The integrated research approach included these methods:

1) household surveys measuring demographic characteristics and socio-economic variables (in all the villages);
2) ethnobotanical formal and informal interviews, including conflict mediation studies involving *jaṅgal* use and management;
3) participant observations, including a two-week field trip up to the *lekh* with six Rai men and 100 sheep, and a three-week trip to Taplejung researching the *chiraito* trade and networking relationships to Tibet and India;
4) a *jaṅgal* species identification list and preservation of samples;
5) a women's health interview;
6) a 24-hour dietary recall;
7) a food frequency checklist;
8) observations of intra-familial distribution of food, "snacking," etc.;
9) a semi-structured survey of *chiraito* collection in all the villages;
10) participatory maps of natural resources identified and drawn by the farmers;
11) interviews and participant observation of *dhāmis* and *lamas*, regarding use of medicinal and incense plants and rituals related to natural resource management;

12) Historical interviews regarding early settlement patterns, origin myths, legends, etc. and

13) Key informant interviews of the farmers perceptions of the current status of the natural resources and the environmental changes which may have occurred.

This information and other techniques of investigating the social organization (kin networks) and economic mode of production, helped to provide a unified picture of the use of *jangal* resources.

In the Sherpa villages the household and key informant interviews were conducted with a bilingual (Nepali and Sherpa) *lama* from Gongtala. In Yangden, the household interviews etc. were conducted with a bilingual Kulunge Rai male (Kulunge Rai and Nepali) who was from Yangden. The womens' interviews were conducted by a Sherpa woman from Gongtala.

Indigenous Knowledge Systems

Indigenous knowledge systems (IKS) is synonymous with folk (local) knowledge being specific to a given culture and important in its adaptation to difficult and unpredictable environments. Alcorn (1995:1) suggested that an inventory, assessment and application of indigenous knowledge for the local management of natural resources is especially needed in "regions of marginal farmland where high-input, capital intensive systems are unprofitable or unsustainable." Etkin (1994) stresses the importance of including 'wild' plants in the assessment because of the indigenous population's dependence on them. Culturally-transmitted ecological knowledge is required to locally manage and use the full diversity of *jangal* resources. Religious beliefs and practices impose cultural restraints on land use systems. Alcorn (1995) noted that only by encouraging farmers to participate in conservation development as actors and teachers will this development be action-oriented and context specific.

The method by which ethnobotanical knowledge is passed on via folklore in non-literate societies is referred to as customary 'scripts' by Alcorn (1995). These customary 'scripts' of indigenous knowledge are intergenerational and have been retained in remote areas, such as the Arun Valley. Nowadays this knowledge is becoming esoteric for specific plants because it is held secretly within certain families. The changing social structure within households arising from marriage and out-migration are contributing factors. An elderly Sherpa lamented, that the indigenous medicine lore he had learned from his father is being lost. His daughters, who are high-altitude herders, were most knowledgeable about the *jangal* ingestibles. They are of marriageable age, and the Sherpa is planning to sell his high-altitude livestock once wedding arrangements are made. Because the woman's affinal relations may reside in another village, the intergenerational knowledge of medicinal plants within this family and their role in traditional healing may become extinct within a single generation.

These indigenous knowledge systems have been used to develop an understanding of the Rai and Sherpa perceptions of the *jañgal* and its present state. Individuals were asked whether in their opinion *jañgal* resources are decreasing or remaining stable. The oral history and ethnic-specific cognitive maps of the *jañgal* and the local perceptions and management of natural resources will be described in Chapter 9.

"Wild plants and minor crops are integrated spatially and temporally into dynamic agricultural systems that exploit and conserve biodiversity as natural resources" (Alcorn 1995:2). Shrestha (1989) asserts that although these 'wild' plants are widely used by farmers as an adaptive strategy for household food security, their indigenous knowledge of collecting, processing and producing various items has yet to be adequately evaluated and promoted. However, according to deBeer and McDermott (1989), a classification of *jañgal* resources is needed, based on who uses them and how their role in the household, local and national economy and the economic value can be measured. It is interesting to note that Alcorn (1989b in Alcorn 1995) showed that the economic benefits from traditional farming which included 'wild' plants and maize swidden fields were higher than individuals who earned wages from employment in towns of the same region. These issues will only be touched upon in this chapter because Chapter 7 will discuss the processing of *jañgal* ingestibles and Chapter 8 will discuss the major *jañgal* resources that are bartered and/ or traded.

Jañgal Resource Collection and Use

When I asked the villagers 'What do you collect from the *jañgal* besides fuelwood?,' they expressed surprise that I was interested in their indigenous *jañgal* knowledge. The *jañgal* carries a stigma among members of the higher economic strata and high-caste ethnic groups. The villagers responded enthusiastically to my questioning by bringing various 'wild' plants to my bamboo hut.

The scale of the local people's indigenous knowledge is reflected in the more than seventy-six *jañgal* resources collected and identified to date according to species, genus, family and local names: Nepali (N), Rai (R) and Sherpa (S) (see Appendix). The importance of these resources is exemplified by their multi-functional role in the rural households. Forty-seven *jañgal* resources are consumed by the local population, thirty-eight are eaten by livestock, nineteen have medicinal value, five serve religious and ceremonial purposes, eleven are used as household implements or building materials, and eleven are traded (both raw state and processed). These are shown in Table 14. This list is not exhaustive and many more *jañgal* resources are available, though not used by the villagers at the present time. Clearly the *jañgal* supplies a vast number of resources.

Table 14
Jaṅgal resource multi-use

Species	Product						
	fod	fo	med	HH	rit	b/t	oth
Allium hypsistum	-	x	x	-	-	-	-
Amomum aromaticum	-	x	x	-	-	x	-
Arisaema flavum	-	x	x	-	-	-	I
Artemisia vulgaris	x	-	-	-	x	-	Y
Arundinaria aristatla	x	x	-	x	-	x	-
A. falcata	x	x	-	x	-	x	-
Berginia purpurascens	x	x	x	-	-	x	-
Bistorta spp.	x	x	-	-	-	-	-
Bombay spp.	x	-	-	x	-	x	-
Brassaiopsis schefflera	x	-	-	-	-	-	-
Cardamine L. hirsuta	x	x	-	-	-	-	-
Castanopsis hystryx	x	x	-	x	-	-	F
Cinnamomum tamala	x	x	-	-	-	-	-
Dalbergia pinnata	x	-	-	-	-	-	-
Daphne bholua	-	-	-	x	-	x	P
Dendrocalamus hamiltonii	-	-	-	x	-	-	-
Dryopteris cochleata	x	x	-	-	-	-	-
Dryopteris fillix	x	-	-	-	-	-	-
Elephantopus scaber L.	x	-	-	-	-	-	-
Elsholtzia blanda	-	-	-	-	x	-	-
Ficus nemoralis	x	-	-	-	-	-	F
F. roxborghii	x	-	-	-	-	-	F
Fragaria vesca L.	x	x	-	-	-	-	-
Gentianaceae	-	-	x	-	-	-	-
Girardinia diversifolia	x	x	-	x	-	x	-
Holboellia latifolia	x	x	-	-	-	-	-
Juniperus incurva	-	-	-	-	x	-	F
Kobresia nepulensis	x	-	-	-	-	-	-
Litsea citrata	-	x	x	-	-	-	-
Musa paradisiaca L.	-	x	-	-	-	-	-
Mussaenda frondosa	-	x	-	-	-	-	F
Myrica esculenta	x	x	-	-	-	-	-
Perilla frutescens	-	x	-	-	-	-	-
Persea odoratissima	x	x	-	-	-	-	-
Phyllanthus emblica L.	x	x	x	-	-	-	-
Phytolacca acinosa	-	x	-	-	-	-	-
Picrorhiza scrophulariiflora	-	x	x	-	-	-	-
Potentilla peduncularis	x	-	-	-	-	-	-

contd.....

Species	Product						
	fod	fo	med	HH	rit	b/t	oth
Quercus lamellosa	x	-	-	x	-	-	F
Rheum australe	-	x	x	-	-	-	-
R. webbianum	-	x	x	-	-	x	-
Rhododendron anthropogon	-	x	-	-	x	x	-
R.setosum	-	-	-	-	x	-	-
Rubus niveus	x	x	-	-	-	-	-
Saussurea gossypiphora	-	-	x	-	-	-	-
Strobilanthes goldfussia	x	-	-	-	-	-	-
Swertia chirata	-	-	x	-	-	x	-
S. multicaulis	-	-	x	-	-	-	-
Symplocos sp.	-	x	-	-	-	-	-
Thysanolaena maxima	*x*	-	-	x	-	-	-
Trewia nudiflora	x	x	-	x	-	-	-
Umbelliferae	-	x	x	-	-	-	I,O
Vaccinium sp.;	-	x	x	-	-	-	-
Zanthoxylum armatum DC	-	x	x	-	-	-	-
Aktingok (S)	-	x	-	-	-	-	-
Arupate	-	x	-	-	-	-	F
Bhurmang sāg	x	x	-	-	-	-	-
Duti Cheru	x	-	-	-	-	-	-
Indrini	-	x	-	-	-	-	-
Jiblolim (R, N)	x	-	-	-	-	-	-
Jurpala (R)	-	x	x	-	-	-	-
Kanchirna	x	-	-	-	-	-	-
Kejo (R, S)	-	-	x	-	-	-	-
Lara	x	x	-	x	-	-	-
Nakshe	x	x	-	-	-	-	-
Pamugoen (S))	-	x	-	-	-	-	-
Phararausi (R)	-	x	x	-	-	-	-
Pipiho (N, R)	x	-	-	-	-	-	-
Salesi (R)	-	x	-	-	-	-	-
Samshipo (N,R)	x	-	-	-	-	-	-
Wabermang (R,N) fo	x	-	-	-	-	-	-
Waldekpa (S)	-	x	-	-	-	-	-
Mushrooms (four types)	-	x (4)	-	-	-	x (1)	-

Key: fod= fodder, fo= food, med= medicine, HH= household items (building materials etc.), rit= ritual, b/t= bartered or traded, oth= other. Code (oth*): Y= yeast starter, I= insecticide, F= fuelwood, O= ornamental, P= paper, W = wormicide.

Among Sherpas and Rais, the same species fills a number of different roles. For example, *toho*, (*Arisaema flavum*) is a plant with a yellow tuber found in the *jaṅgal*.

The tuber is eaten by bears and wild boars, used to make *raksī*, to treat stomach ailments, to make bread or is added to soup. During the monsoon season, its leaves even serve as a rain hat. Sometimes the farmers do not know the specific reason behind the use of the *jaṅgal* resource and just say, "it's good for you." Various parts of the main species (leaves, flowers, fruits, roots, tubers, seeds) are used (see Appendix).

Some raw materials (plants and animal resources) are used directly and require no processing. 'Wild' fruits are an example because they are consumed on the spot by herders and farmers as they walk through the *jaṅgal*. Tubers require a lengthy processing procedure described in Chapter 7. Woven bamboo mats are another example discussed in Chapter 8.

Social Groups and Actual Use

Besides classifying the plants, I also documented who gathered and used these resources, with what frequency, and during which season. Many medicinal plants are available (nineteen were named), however, only a few are universally used to treat human illnesses. The accessibility of the medicinal roots is a problem because the majority of them are found at the *lekh* which is more than 4,200 m in elevation. The knowledge and usage of these is restricted to herders who go to the *lekh* every monsoon period. *Chiraito (Swertia chirata)*, a medicinal plant, is an exception because it grows at a lower elevation near the home villages and is widely used.

Jaṅgal resource use in the rural community varies by gender, age, ethnicity and economic status. The group which utilizes the particular *jaṅgal* resource is the one which holds knowledge about it.

Most *jaṅgal* ingestibles (food resources) are easily available, especially 'wild' greens, which may be collected within an hour's distance from the households when a household member is collecting firewood or bringing water. *Jaṅgal* food resources are generally, but not exclusively, collected by Rai and Sherpa women. This is not a set rule because bamboo shoots and mushrooms are gathered by family members who stay at the cow shelter (*goṭh*) and are brought to the village when they come for food supplies. The herders are usually children, ten years and older, or a woman, if the household has only small children. Men are also seen carrying *jaṅgal* ingestibles to their households.

Young children who run freely near their houses snack on 'wild' fruits, for example, *asare (Mussaenda frondosa)* and *ainselu* (red raspberries). Even though these fruits are important for their diet, children, nevertheless, are often wrongly dismissed as knowledge holders.

Types of *Jańgal* Resources

Firewood

Local knowledge about firewood collection is not a major focus of this research because it has been covered extensively by Fox (1983 and 1984) and Metz (1990). However, a few important points regarding firewood collection will be highlighted. All the farmers stated that they collected firewood on private or community land because there was no government land in this area. They felt that once the Cadastral Survey was carried out, the *kipat* (communally-held land) would be claimed as government land.

Many species of firewood are available. However, the farmers prefer firewood species which burn better and faster.

Table 15
Preferred firewood species

Local name	Scientific name
Asare	*Lagerstroemia floribunda*
Chimal	*Rhododendron barbatum*
Gobre salla	*Pinus wallichiana*
Katus	*Castanopsis indica*
Kharane	*Symplocos ramasissima*
Khasru	*Quercus semacerpifolia*
Shukpa (S)	*Juniperus incurva*
Utis	*Alnus nepalensis* D.Don

At the high pasture, firewood species are rare, but fortunately *chimal* (*Rhododendron barbatum*), a very plentiful source of firewood, is found there. Because it ignites quickly, it is highly valued, especially when the herders are cold and damp from the monsoon showers. *Shukpa*, also found at the *lekh*, is a favored wood because it burns quickly and helps other firewood to dry faster.

Firewood is usually collected in the early morning hours. Collection takes about one hour per day for the Sherpas, while for the Rais, a round-trip distance of about three hours is required. The Rais have to travel further to fetch firewood because Yangden is situated in a valley and the forests are reduced around it. The firewood supplies last one to five days depending on family size and season, the main determinants of the consumption rate. After the winter months, the farmers collect sufficient firewood to last until the monsoon subsides, and store it on rafters under their houses.

Timber is not sold for cash and the quantity harvested is only for specific subsistence needs. *Gobre sala* wood (*Pinus wallichiana*) is used by Sherpas for construction of dwellings (the roof and building poles). Wood is also used for agricultural implements; mortars and pestles. For small-scale *jaṅgal*-based household enterprises, such as alcohol production using *jaṅgal* ingestibles, firewood is required. Generally only men cut the trees (fresh wood) while women gather dead branches off the ground. Tree lopping for fodder is done by both men and women in the area. Restrictions on cutting fresh wood is enforced by fines imposed by the village council regulations. This pattern of resource management will be further discussed in Chapters 6 and 9.

Ephrosine Daniggelis

A Rai woman with a *doko* of firewood

Fodder

Fodder scarcity occurs for animals during the winter months when the ground is covered with snow. In the Sherpa villages, an average of one to two hours per day is needed to collect a *doko* of fodder, while in Yangden, it take twice as long. Approximately 35-50 *doko* of fodder are needed per week depending on the number of livestock owned by the farmers.

Trees, grasses and agricultural by-products (maize and millet stalks) are sources of fodder. Although there are thirty eight types of fodder listed by Rai and Sherpa farmers, only one-half were identified as preferred sources shown in Table 16.

Table 16
Preferred fodder sources of Rai and Sherpa farmers

Local name	Scientific name	Used by (Rai/Sherpa)
Amlise	Thysanolaena maxima	R
Anibarya		S
Buki	Kobresia nepalensis	S
Chap (N), gok (S)	Vimalis	R, S
Ciula, cyiula		S
Dudhilo	Ficus nemoralis	R, S
Gogane	Saurauria napaulensis	R, S
Habru		S
Kaulo	Machilus gamblei	S
Khaniyo	Ficus cunia	R
Kosela		R
Māliṅgo ghās	Arundinaria aristatla	R, S
Mulapate	Elephantopus scaber L.	R, S
Nibharo	Ficus roxborghii	R, S
Painyu	Prunus cerasoides	S
Bangshirs		S
Payungmalpaiyuma	Persea	S
Ramar (S)	Brassaiopsis schefflera	S
Shingane ghās	Drepanostachyum	S
Thotne	Ficus hispida/ Polygonum molle	R

key: N=Nepali, S=Sherpa, R=Rai.

Besides its known utilitarian and economic importance as fodder and brooms, *Amlise* (*Thysanolaena maxima*) has ecological importance because it is "a vital plant resource for erosion control" (APROSC 1991:5). Another example of a mismatch of scientific and local knowledge can be exemplified by *Elephantopus scaber* L. which according to Mabberley (1993:202) "is a bad weed in warm regions." For the Rais and Sherpas, however, it is a source of cow and sheep fodder. The farmers named fodder sources which were mutually exclusive for goats, cows and high-altitude animals.

In the Apsuwa Valley, the transhumance herding system exploits different altitudinal zones on a year-round basis. This strategy allows sufficient fodder for the animals, since high-altitude animals are not in competition with door animals because different niches are grazed. Although forty fodder species have already been named, there are still many yet to be identified in the Apsuwa Valley.

A variety of terms are used to describe how the fodder affects animal health and milk production. Fodder types were noted as "nutritious," "milk producers" and "very good." Several fodder species have been planted along the borders of agricultural fields, saving time and ensuring a constant supply of feed for the animals, especially during the scarce winter season.

Ingestible Resources
Floral
The term ingestible plants encompasses both 'food' and 'medicinal' plants because the same plant often has a dual function (Etkin and Ross 1991:231). For example, Golden raspberries (*Rubus ellipticus*) which grow wild around the villages are eaten as a fruit and the juice is useful to cure fever and treat coughs (Manandhar 1989a). The role of *jangal* ingestibles provides important nutrients, as well as flavor and variety to the local diet. Gopalan et al (1984:37) remarked that garlic, besides its use as a spice and flavoring, "is believed to contain active principles which inhibit the growth of putrefactive bacteria in the intestinal tract." And *Dioscorea bulbifera* treats sore throats (Bremness 1994). They are eaten during the cold winter months when sore throats are a frequent occurrence.

In Nepal, different ethnic groups often cognize the wild/domestic continuum (Etkin (1994) in their own distinctive ways. Whether a given resource is viewed as 'wild' or domestic may influence how it is valued, regardless of its nutritional value. Sherpas who are well-off economically (sufficient land and/or livestock) consider *jangal* foods low status and 'poor person's food.' They apologized to me whenever they served *sisnu* or other *jangal* foods, even though I told them that I relished these foods. *Sisnu* is the Nepali name used interchangeably for three similar types of 'wild' leafy greens (Figure 4): stinging nettle (*Urtica dioica*); Himalayan nettle (*Girardinia diversifolia*); and *Girardiana palmata*. All three types of *sisnu* are plentiful in the *jangal*. Whenever *sisnu* is used throughout the text it will generally refer to these species, unless the common names, stinging nettle or Himalayan nettle, are noted.

Figure 4. Three species of *sisnu*. Left: *Girardiana palmata,* Wild Edible Plants of Nepal (1982:134). Center: *Urtica dioica* (Stinging nettle), Wild Edible Plants of Nepal (1982:262). Right: *Girardinia diversifolia,* (Himalayan nettle). Manandhar 1989:46.

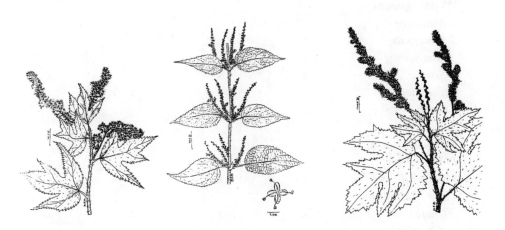

There is a seasonal variation in the availability of *jaṅgal* ingestibles (food sources) with more varieties available during the monsoon. Table 17 shows the *jaṅgal* ingestibles eaten during the monsoon period according to the household survey.

Personal observations revealed additional *jaṅgal* ingestibles used by the farmers which were not mentioned on the household survey. This is the reason the research design used many techniques in order to 'capture' the range of 'wild' foods collected. When I accompanied the Rai shepherds to the *lekh* during the monsoon, I saw them gathering *mangan sāg* to eat as an accompaniment to their main staple. Bundles of the 'wild' greens were also carried back to their home village by men who had portered food supplies to the summer pastures. Not all of the numerous varieties of mushrooms that are consumed were identified. These mushrooms grow on fallen trees, rocks and on the grass. The *jaṅgal* ingestibles in Table 18 were eaten during the winter period, according to the household survey.

Table 17
Jaṅgal ingestibles collected during the monsoon

Jaṅgal Foods	Gongtala	Dobatak	Yangden	Saisima
'Wild' greens				
- Bhurmang			x	
- Jaringo sāg	x	x		
- Mangan sāg	x			
- Sisnu	x	x	x	x
Niguro (ferns)	x	x	x	x
Fruits				
Tau	x			
Siltimbur (con*)	x			
Phi to (tuber)	x		x	x
Theki phul	x	x		
Tusa (bamboo)				
- Ghure	x	x	x	x
- Kālo nigālo	x	x	x	x
- Tite nigālo				x
Mushrooms				
- Kharne	x	x	x	x
- Cyitle	x	x	x	x
- Kālo (black)	x	x	x	x
- Rato (red)	x	x	x	x
- Cilme				
Cilding shamu	x	x	x	x

* = Also a condiment.

Table 18
Jaṅgal ingestibles collected during the winter months

	Gongtala	Dobatak	Yangden	Saisima
Sisnu	x	x	x	
Niguro	x			
Bhurmang sāg	x		x	
Nam chharog (fruit)*	x			
Mashita (S) (spice)	x			
Ban tarul (tuber)			x	
Gitta dayagur (tuber)			x	

* Phyllanthus Emblica L.

Rai shepherd carrying *mangan sāg (Cardamine hirsuta)*

Table 18 reveals that tubers are only eaten by Rais during the winter months. They are used as a staple substitute for this period which coincides with *anikāl* (food shortage time).

Etkin (1994) asserts that despite a conceptual distinction between 'wild' and domestic plants which can be culturally important, in some cases, it is difficult to differentiate between 'wild' and semi-domesticated plants because the knowledge varies culturally and temporally. For example, *danur* (S), *jaringo sāg (Phytolacca acinosa* Roxb.) is an example of a semi-domesticated *jaṅgal* ingestible. *Danur* was planted as seeds from the *jaṅgal* seven years ago and then continues to grow for many years. It ripens from mid-April to September. *Danur* is made into a spiced vegetable soup which accompanies the main staple meal.

Timur (Zanthoxylum armatum DC) is a cultivated shrub or small tree about three meters tall which grows in the temperate region. *Timur* will grow for many years, so it needs to be planted only occasionally. The tree takes three years to bear its red seeds. The fruit is used as a spice, in Sherpa stew, *shakpa* (S) and flavors *sisnu* served as a side dish and the fermented cheese soup (*somar;* T, *so.mar.ra*). The Sherpas only use the seeds for home consumption. However, *timur* harvested from other areas is marketed in India. The seeds are used to aid in digestion, to relieve colic and treat cholera. The oil is used in perfume, medicinal powder and germicides (Manandhar 1980).

Kachu (T), (*gyorel*), (T, S) is a semi-domesticated 'wild' tuber found in Saisima. The farmers said they plant it only one time for it to regenerate. The root, when it is young, is made into a type of sweet. *Kachu* can be cooked and eaten like a potato or even fermented and prepared as a beverage, for example, *raksī* (distilled alcohol). The leaves are a source of fodder for cows.

'Wild' yam (*Bhyākur; Dioscorea deltoidea wall*) is a climber vine found in the *jaṅgal* near Yangden, and also in Bhutan. A vine can yield up to ten kilograms of tubers (Subba 1996). There is no gender division of labor in digging up the 'wild' yams, which act as an important staple for the Rais during the lean season. The bulbs on the vine are bitter and only eaten under dire emergency conditions (Subba 1996). 'Wild' yams have extra health-mediating potential (Etkin 1994) because the tubers are rich in diosgenin (glycoside), a precursor of progesterone used to make contraceptive pills, and contain cortisone and steroids, anti-inflammatories to treat rheumatism (Bremness 1994; Mabberley 1993). In 1991/92 in Nepal, eight tons were collected, bringing in a revenue of Rs 22,000. *Dioscorea* are cultivated on large plantations in India near the Bhutanese border, where the annual revenue is US$ 319,000 (Subba 1996).

Ramba (*Allium hypsistum*) are 'wild' thin greens gathered around the Sherpa houses. The planting of *ramba* is not necessary because it is self-generating. They are pounded with chili and salt to make a side dish served with boiled potatoes or other foods. Although only eaten by Sherpas, *ramba*, referred to as 'wild' garlic by Pohle (1990), is reported to be used for flavoring the lentil soup (*dāl*) in Nepalese villages.

Toho (*Arisaema flavum*) requires a long trip to the *lekh* to acquire, a characteristic of famine foods. It is eaten only once or twice during the monsoon season to add variety to the local diet. Sacherer (1979a) suggests that the Sherpa's long history of consumption of this "potato-like root" (*Arisaema intermedium Bl.*) led to the rapid adoption of the potato. *Arisaema flavum* is a variety of this plant. These tubers require a lengthy processing because they have allelochemicals (secondary compounds) produced by plant tissues and/or by associated microflora which act as defenses and inhibitors to competitors (Huss-Ashmore and Johnston 1994).

Non-Floral

The Rais are noted for fishing in the Apsuwa *Kholā*. The *asala* (snow trout) is considered the most delicious fish found in the Apsuwa *Kholā*. The Rais present the fish as gifts to wealthy Sherpas.

Honey and the honeycomb are another example of animal products collected from the *jaṅgal* for food. Honey is produced by *Apis dorsata*, referred to as the 'rock bee.' The large comb is found on the underside of the steep rock cliffs located high in the *jaṅgal* above Saisima. Obtaining the honey requires skill and is very difficult because the person must climb up on a high rock to reach it. Bijale, a Yangden Rai *dhāmī*, was the only person known who could gather the honey. He passed away unexpectedly during my stay in the village and the knowledge was lost. Although I interviewed his son, I was unable to obtain more explicit information regarding the collection process. A single comb of honey may yield up to 3.5 kilograms (Negi 1992).

The honey has an extremely rich taste and is believed to have medicinal properties. If a person eats "too" much honey, it is said he gets into a drunken stupor and may vomit. Honey has limited availability, partially due to its inaccessibility. The larvae of the bee are eaten immediately. The honey-filled comb is cooked up or eaten and chewed before discarding the wax. I ate the larvae in Dobatak, which had been fried in butter.

Historically, hunting has provided an important protein source for villagers who experience "meat hunger." Hunting is now prohibited within the Makalu-Barun National Park and Conservation Area in order to protect the existing wildlife. Poachers from neighboring villages come to this area to hunt wild boar. During the monsoon season, Rai men with rifles who were not wary of the repercussions of the deities residing in the *jaṅgal*, came from Yachamkha village, a two-day walk from Gongtala. They went to hunt *banel* (wild boar) in the Buddhist sacred forest near Saisima. It was raining hard and the hunt was unsuccessful. The villagers said the reason they had no luck was because they angered the deities who caused the rain to protect the *banel*.

Medicinal Plants (*Ja ibuti*)

The sub-alpine and alpine zones are a rich source of medicinal plants. Collection of these plants in the alpine zone is usually done by herders during transhumance. More than four hundred medicinal plants are known in Eastern Nepal, of which only about twenty are exploited for commercial purposes (Malla and Sakya 1984-5 in Shrestha 1989). The export figures on the status of alternative forest products do not reflect the actual quantity of trade, because a large quantity of medicinal plants are smuggled out of mountain areas to avoid royalty payments

(Aryal 1993: Yonzon 1993). Yonzon (1993) further asserts that uncontrolled collection of medicinals can lead to extinction because more than half of those extracted from Langtang National Park are destroyed in their entirety. The largest volume of medicinals goes to India rather than directly overseas. The processing of medicinal herbs in Nepal is handled by Herbs Production and Processing Company Ltd (established 1981) in conjunction with Royal Drugs, the largest consumer in Nepal. Gorkha Ayurveda Company and Dabeer Company which are privately-owned also process medicinal herbs.

Conservation of medicinals occurs in sacred forests (*Devī Than*) where their extraction is prohibited. These sacred forests are a rich genetic storehouse. The amount of medicinals collected at the *lekh* is minimal with the exception of *chiraito* (*Swertia chirata*), found in grassy areas at lower elevations, which is traded on a large scale. The farmers rely on medicinals as their main source of home remedies to treat illnesses, as shown in Table 19.

Table 19

The local use of medicinals by ethnic groups

Plant name	Use	Groups Using
Allium hypsistum	cuts	Sherpa
Amomum aromaticum	colds, cough	Sherpa
Arisaema flavum	stomach ailments	Sherpa
Berginia purpurascens	severe diarrhea, after childbirth	Sherpa
Gentianaceae	colds and fever	Sherpa
	cuts	Rai
Litsea citrata	severe diarrhea	Rai, Sherpa
Phyllanthus emblica L.	diarrhea, stomach ailments	Rai
Picrorhiza scrophulariiflora	cold, stomach ailments fever, headaches	Rai, Sherpa
Rheum australe	fever, cuts, body ailments	Rai, Sherpa
R. webbianum	fever, cuts	Sherpa
Saussurea gossypiphora	cuts	Rai, Sherpa
Swertia chirata	cold, cough, stomach ailments	Rai, Sherpa
S. multicaulis	cuts	Rai, Sherpa
Umbelliferae	fever	Rai, Sherpa
Vaccinium sp.	fever	Sherpa
Zanthoxylum armatum	stomach ailments	Sherpa
Qurkim (S)	stomach ailments	Sherpa
Kejo (S)	wound, stomach	Rai, Sherpa
Kalinakchun (S)	fever	Sherpa

More medicinal plants are used by Sherpas than Rais, eighteen for the former to ten by the latter. The uses of these medicinals are localized to the area, and knowledge is held by specific farmers. Although the ritual knowledge is secretly held within families, the medicine is universally shared. Rai farmers who were suffering from various ailments would come to the household of a Sherpa who was well known for his knowledge and get a supply of medicine. They would bring *raksī* in lieu of a payment.

The *jańgal* ingestible *chiraito* is the most common source of indigenous medicine for both Rais and Sherpas. According to Negi (1992), *chiraito* has anti-helmintic properties, very important because ascaris is endemic among the villagers in my research area. *Lamas* also use medicinals as a part of their healing ritual. The Saisima *lama* gave a foreigner *siltimur* (*Litsea citrata*) for his stomach ache.

The preparation of the medicinal roots, for example, *kutki* (*Picrorhiza scrophulariiflora*) usually involves the following procedure: first the root is softened in hot water, then pounded in a mortar, and afterwards, the solution is cooled. Then the sick person drinks it.

Medicinal plants are also used to treat livestock, especially important in areas where veterinary services are unavailable. The *Keju* (*S*) (*Rheum webbianum*) root, was cut into pieces and boiled in water. The mixture was then given to a cow which had fallen and hurt its leg. *Karapi jar* is a medicinal plant used to treat cuts or wounds of both humans and livestock. *Siltimur* is also given to domestic animals by Sherpas.

Poisonous Plants (*Bikh*)

Bikh (*Aconitum spicatum*), is a group of plants commonly referred to as monkshood. *Bikh* is considered as a separate category because of the cultural implications of its use, which will be discussed in Chapter 6. *Bikh* roots are *jańgal* ingestibles which contain highly toxic alkaloids (*aconitin*). They are found in the alpine and sub-alpine zones of the Himalaya. There are seventeen or eighteen types of *bikh* in Nepal and many are not yet identified according to Shrestha (in Ramble 1993). *Monkshood* tubers (*Aconitum napellus*) were found for sale in the upper Arun by Shrestha (1989).

Aconitum napellus changes to *aconite*, in the form of a bitter powder. It has long been used as a medical treatment for leprosy, cholera and rheumatism. It has to be administered in controlled doses; otherwise it is likely to cause dangerous side-effects because it is a poisonous drug. This drug also treats fever, post-fever weakness, diarrhea, dysentery and is externally applied for diseases such as neuralgia (Negi 1992).

Bremness (1994:137) points out that "the tuber is one of the plant kingdom's most powerful nerve poisons." The roots have a slight taste, but consumption is followed by a persistent sensation of tingling of mouth, tongue and stomach. When an overdose has been taken, it can spread throughout the body and result in numbness. Fatalities result from cardiac arrhythmias. In case of poisoning, the stomach needs to be washed out with emetics. *Aconite* is considered one of the most deadly poisons and its use is now restricted (Wealth of India 1985).

Bikhma or *seto bikuma* (*Aconitum palmatum*) is used by farmers as an antidote to the *bikh* poisoning. Its tubers contains *vakognavine* (sucrose). *Bikhma* is also used as an application, and to treat diarrhea and rheumatism (Malla 1982).

Jangal Non-Ingestible Resources

Utilitarian: Bamboo
Bamboo will be considered as a separate category because of its cultural importance for Rai households. Rajbhandari (1991:17) labeled Rai culture as a "bamboo culture" because bamboo is the resource most heavily used (see Seeland 1985).

For the marginalized Rai farmer, "bamboo is life." The importance of bamboo can be traced back to ancient times, and is found in the creation myth of the Rais (see Chapter 6). When I first stood on a hill overlooking Yangden, I viewed a scattering of thatched bamboo houses among golden maize fields. The entire Rai house (walls, roof and platform) is made of various bamboo species. Pillars and roof beams are bamboo and bamboo mats are intricately and tightly woven to protect the family against wind and rain. Women and children are seen carrying large bamboo containers filled with water. The water is collected from split bamboo pipes placed above the river, and is carried on their backs in *dokos*. Babies are kept in a *kokro* (woven basket). Buttermilk, butter, *raksī* and *jād* are stored in bamboo containers of various sizes. Bamboo leaves supply fodder for the young animals kept by the households. Animals wear bamboo muzzles as they are led through maturing fields. It is a common sight to see Rai men sitting outdoors on the bamboo house platform weaving *bhakāri* (cow shelter bamboo mats), *namlo* (bamboo straps for carrying goods) or *ghum* (bamboo raincoats). The raincoats are a very adequate rain repellent. Up at the *lekh* with the shepherds, bamboo musical instruments added entertainment to the long hours spent around the fire. Even the slippery bridges that I crossed were often large bamboo poles fastened together with bamboo lashings. Rat traps (*darama*) are made out of *mālingo* (*Arundinaria aristatla*) and hidden amongst the stored maize. Bamboo is made into toys. Young boys made compressed air guns out of *mālingo* with berries as the ammunition. Bamboo is everywhere and used by both genders and all social groups in Yangden.

Bamboo belongs to the *Gramineae* family and is abundant as an understory in moist temperate forests. It favors disturbed areas where there is a light to moderate canopy of deciduous species (Rajbhandari 1991; Sharma 1982). Bamboo grows and regenerates quickly and responds well to cultivation. In the Apsuwa Valley, two general types of bamboo grow, a large-statured bamboo and a number of smaller bamboos. These types show altitudinal differentiation. The larger bamboo (*bāns*) is found at less than 2,000 m elevation. Its culms are more than eight meters high. This bamboo is cultivated on private land. The small bamboos grow naturally in the *jaṅgal* on shaded slopes, swidden fields and pasture areas. They grow up to seven meters high and their diameter is smaller than the *bāns*.

Children also make music.

Table 20
Small-statured bamboo species

Scientific name	Local name	Uses
Arundinaria falcata	*Kālo nigālo*	fodder, head strap food, raincoats, baskets
A. aristatla	*Mālingo*	house building, food basket products, fodder
Drepanostychyum intermedium	*Tite nigālo*	mats, food, baskets and fodder
Bombay spp.	*Singane*	baskets

The vast number of products made out of bamboo shows what a versatile material it is. Its strength, straightness, hollowness, lightness and rapid growth are the features which account for this (deBeer and McDermott 1989).

Utilitarian: Fibers

Allo (*Girardinia diversifolia*), a fiber plant, grows in the temperate zone at an altitude of 1,200 to 3,000 m. Formerly, *allo* was an important plant in the area for homemade clothes. The leaves are eaten when young. When fully matured, the plant is used by Rais for rope, string and gunny sacks. The inner bark produces a strong and very long fiber which is used in cloth-making. Brooms are made from *amriso* (*Thysanolaena maxima* [*Roxb.*]).

Culturally-Important Artifacts

Kharnang (T), *binayo* (N) is a mouth harp usually made by Rais from *mālingo*. *Lumu* (S), *bāsuri* (N) is made of *māliṅgo*.

Ornamental Uses

Poshi (T, *sposhil*) are collected from the *shukpa* (*Juniperus incurva*) at the *lekh*. The seeds are used to make necklaces or bracelets by Sherpa men and women while they tend the animals.

Income Sources

Most products made out of *jaṅgal* resources are used by the household, bartered or sold in local markets. These products are more accessible to the poorer population. Selling woven bamboo products is a minor source of income for the Rais, but very important because few alternative income-earning opportunities are available. These products are labor-intensive, and are usually made during the slack season, while tending animals, or by the elderly who are home-bound.

The most valuable commercial *jaṅgal* resource in this area is *chiraito* (*Swertia chirata*) which will be discussed in Chapter 8. *Chiraito* brings high income and involvement in regional and international trade linkages.

Bhakāri (Livestock Shelter Roof Mats)

Bhakāri are *māliṅgo* roof mats woven by skilled Rai craftsmen. Young men prepare the *māliṅgo* for the older men, an example of gender and social differentiation of labor activities. The *māliṅgo* is picked above Yangden and Gongtala. One *bhakāri* requires forty to fifty stems of *māliṅgo*. I always knew when *māliṅgo* was being carried from the *jaṅgal* because of the loud shattering sound. The men ran down the hillside with the long *māliṅgo* trailing over their shoulders. The *māliṅgo* is placed above the fire on racks to dry for seven months. The process of making *bhakāri* is labor intensive. For this reason, it is usually elderly Rai men who are seen weaving them all day outside their homes .

A *bhakāri* sells for only Rs 30, very little in comparison to daily trekking wages, which can range from Rs 80-150. They are sold in the nearby Rai villages of Waleng, Chhoyeng and Yaphu. *Bhakāri* are also bartered. The high-altitude animal shelter (*goṭh*) is made of seven *bhakāri* (5 x 3 m each) placed over a frame of wooden sticks. The *bhakāri* are replaced yearly because they rot quickly during the damp monsoon. Heat and smoke retards the decay of these *bhakāri*. Metz (1990) found animal shelter construction to be the fourth largest use of forest biomass. Due to exploitation, bamboo is communally-managed. One Rai farmer said that because pressure on the *jaṅgal* for *māliṅgo* has resulted in scarcity, a few Rais have migrated to Sikkim looking for work. The communal management of *māliṅgo* will be discussed in greater detail in Chapter 9.

Only Yangden Rai men make the various woven baskets and mats. The women say the work is very difficult and they are busy with other household tasks, as well as weaving *rāri* (woolen blankets) and purses.

Lokta (Paper Bark Bush)

Lokta (*Daphne bholua*), a paper bark bush, grows as an understory shrub between 1,500 to 3,000 m in coniferous and broad-leafed temperate zone forests. The plant is found in the *jaṅgal* and grows one to three meters high. The plant may be "coppiced" just above the ground, which allows for regeneration. *Lokta* may be promoted by partial disturbances of temperate forests of the Arun Basin. The inner bark is dried and generally sold directly to middle men who transport it to small paper-making enterprises where it is made into hand-made paper. It also provides a local source of income for the farmers.

A Rai man weaving a *bhakāri*

Usually Rais or Tamangs come from neighboring villages and inform the local Sherpas or Rais of the current market price for those products which are in demand in Hile, a market town four to five days' walk to the south. If it seems profitable, during the agricultural slack period a family member, whoever is free, will cut the *lokta*. In Yangden, *lokta* is seen drying on rafters in the sun, which saves firewood. They are extracted in September or from March to June and later traded. In 1992, the price for one kilo of *lokta* was Rs 7. Khatri (1994) noted that expanding *lokta* cottage industries provides direct employment for 1,500 families, with 100,000 tons produced in 1984. In 1984, a national study by the Department of Forests revealed that Sankhuwasabha District provided 14.1 percent of the total gross stock of *lokta* for Nepal (Bhadra et al 1991).

Dokos (woven open baskets) are made by Yangden Rais and sold to Sherpas in Gongtala, Dobatak and Saisima and Rais in Yaphu, Chhoyeng, Tamkhu and Mengtewa. The *ghum* (raincoat) is made of *māliṅgo* by Yangden Rais and bought by Sherpas and Rais. *Dalo* is a woven basket with a lid made out of *kālo nigālo* (*Arundinaria falcata*). It is used to carry the shepherds' food supplies and bedding to the *lekh* during the summer monsoon season. They are made by Rais of Yangden and sold to Sherpas for Rs 100. Baby baskets (*kokro*) woven from *māliṅgo* by both Rais and Sherpas sell for Rs 100. Hair combs (*kange*) made by Yangden Rais of *māliṅgo* and *pāteliṅgā* wood sell for five rupees.

Kange and bamboo container for butter

Religious and Ceremonial Purposes

Plants are used by both Rais and Sherpas for religious and ceremonial purposes, as shown in Table 21.

Table 21

Religious plants used by Rais and Sherpas

Local Name	Botanical Name	Groups Using
Dupi	Juniperus incurva	Rais, Sherpas
Khobal po	Michelia champaca	Rais
Kisur (S)	Rhododendron setosum	Sherpas
Purmang (S)	Elsholtzia blanda	Sherpas
Seula (R)	Castanopsis hystryx	Rais
Sunpati	Rhododendron anthropogon	Rais, Sherpas
Titepāti	Artemisia vulgaris	Rais, Sherpas
Dhupi	Cryptomeria japonica	Rais

Both Rais and poorer Sherpas collect *sunpati* (*Rhododendron anthropogon*) at the *lekh* during the summer transhumance. One basket of *sunpati* is bartered with the Yaphu Rai for eight *pāthi* (32 kg) of maize. The round trip journey takes three to four days, and *sunpati* is usually collected by shepherds on their return from the *lekh* after delivering supplies. The *sunpati* is used for offerings in deity *pūjās* and also by *lamas* in their rituals.

Dhupi (*Juniperus incurva*) grows in the alpine zone. *Juniper* is made into a powdered pulp and used as a major ingredient in incense for Buddhist rituals. *Tashing*, (S) (*Abies spectabilis*) is erected in the courtyard of Buddhist houses after a baby is born.

Wild animals, feathers and plants are used for ceremonies and rituals, dance costumes and accessories by traditional healers (*dhāmīs*). *Dānphe* (*Impeyan Himalayan Monāl*), a type of wild pheasant, is used by the Rai *dhāmīs* for their headdress.

Figure 5 shows the multi-use and local knowledge of the *jaṅgal*. Figure 5 was prepared for a workshop on Regenerative Agricultural Technologies for the Hill Farmers of Nepal (NERRA/IIRR 1992).

Figure 5. The Multi-Use and Local Knowledge of the *Jaṅgal*

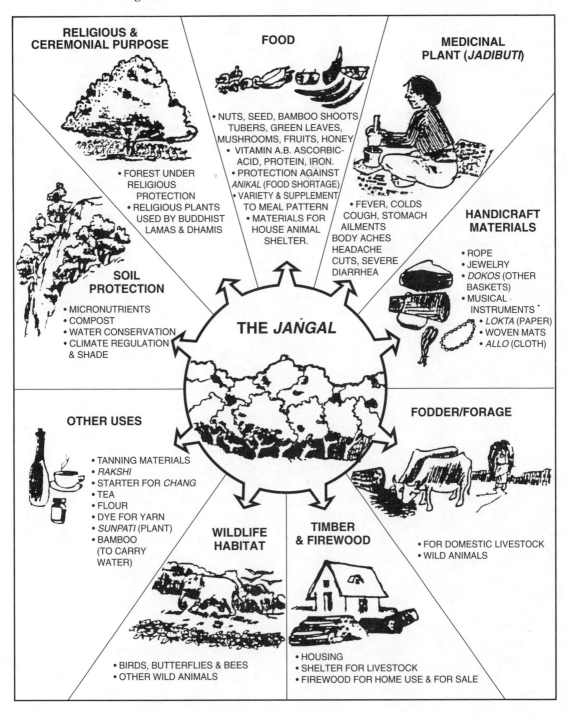

RELIGIOUS & CEREMONIAL PURPOSE

• FOREST UNDER RELIGIOUS PROTECTION
• RELIGIOUS PLANTS USED BY BUDDHIST LAMAS & DHAMIS

FOOD

• NUTS, SEED, BAMBOO SHOOTS, TUBERS, GREEN LEAVES, MUSHROOMS, FRUITS, HONEY
• VITAMIN A.B. ASCORBIC-ACID, PROTEIN, IRON.
• PROTECTION AGAINST *ANIKAL* (FOOD SHORTAGE)
• VARIETY & SUPPLEMENT TO MEAL PATTERN
• MATERIALS FOR HOUSE ANIMAL SHELTER.

MEDICINAL PLANT (*JADIBUTI*)

• FEVER, COLDS COUGH, STOMACH AILMENTS BODY ACHES HEADACHE CUTS, SEVERE DIARRHEA

SOIL PROTECTION

• MICRONUTRIENTS
• COMPOST
• WATER CONSERVATION
• CLIMATE REGULATION & SHADE

THE *JAṄGAL*

HANDICRAFT MATERIALS

• ROPE
• JEWELRY
• *DOKOS* (OTHER BASKETS)
• MUSICAL INSTRUMENTS
• *LOKTA* (PAPER)
• WOVEN MATS
• *ALLO* (CLOTH)

OTHER USES

• TANNING MATERIALS
• *RAKSHI*
• STARTER FOR *CHANG*
• TEA
• FLOUR
• DYE FOR YARN
• *SUNPATI* (PLANT)
• BAMBOO (TO CARRY WATER)

FODDER/FORAGE

• FOR DOMESTIC LIVESTOCK
• WILD ANIMALS

WILDLIFE HABITAT

• BIRDS, BUTTERFLIES & BEES
• OTHER WILD ANIMALS

TIMBER & FIREWOOD

• HOUSING
• SHELTER FOR LIVESTOCK
• FIREWOOD FOR HOME USE & FOR SALE

Figure 5: The Multi-Use and Local Knowledge of the Jangal.

CHAPTER 6

Religious and Cultural Significance

The *jaṅgal* provides important cultural and religious symbols for the Rais and Sherpas living on its fringes. The farmer's symbolic relationship with the *jaṅgal* is laden with cultural meanings and woven into their oral tradition, that is, origin myths, legends and rituals. In non-literate societies such as the Kulunge Rais, indigenous oral tradition can be used to describe the cosmological system. Gaenszle (1993:198) discusses how the Mewahang Rai's indigenous oral tradition, known as *Muddum*(M), shows the Mewahang social order established by the ancestors and their traditional way of life. Mythology is one symbolic idiom "through which deities and other ancestral beings are conceived" and interpreted by the populace.

The Origin Myth

If we look at the mythical origin for a view of how Rais see themselves (Gaenszle 1993), the *jaṅgal* plays a prominent role in their ethnic identity. The origin myth relates the Rai creation to plants and animals. There are many references to the *jaṅgal* and nature interwoven within the myth itself. These references exemplify the spiritual and ecological importance of the *jaṅgal* for the Kulunge Rais.

> The primeval snake, referred to as the *nāgi*, is considered the first being and from which originated all creation. It is said to reside in the rivers, especially at the source up in the mountains. The *nāgi* is pierced by a certain straw, *siru* which gives birth to the Primeval Mother....Later Somnima, a great granddaughter gives birth to the first thorny creeper, *smilax;*[25] and all other varieties of plants, to the two kinds of bamboo, *mal bās*, *Bambusa nutans*; and *gope*, *Cephalostachyum capitatum*; who are called Lalahang and Pakpahang-, and finally to Tiger, Bear, Monkey and the First Man, Tumna (in this order) (Gaenszle 1993:200). Tumno, last to be born, came into the world carrying a bow and arrow (Ramble and Chapagain 1990:31).

In this origin myth, bamboo is "born" before the wild animals and man. In Chapter 5, I discussed that bamboo is the backbone of Rai culture and as shown by

---

25 The plant stem is used to make the curved drumstick of the shaman (Ramble and Chapagain 1990).

the origin myth, its ritual importance has been maintained up to the present. Ramble and Chapagain (1990:31) remark that the sequence of birth order (of animals and man), marked by progressively longer digits, closely approximates the evolutionary order.

Further on in the origin myth more "mundane" and human founders of Rai culture and settlement are mentioned.

> The widely known culture hero Khakculukpa, who is an orphan, was left by his two sisters as a small boy to live alone in the forest, residing on top of a banana tree and eating the fruits of the jungle. He is harassed by the persistently evil forest witch (the *ban devī*, Yagangma), but saves himself through his trickery and finally defeats her. Later on he begins to practice agriculture and fishing..... So Khakculukpa is the classical culture-bringing hero, who passes through a typical life-crisis while coming of age. But through vanquishing the forces of wilderness he establishes the foundation of Rai-culture: agriculture and hunting, marriage by theft, housebuilding and finally an ambiguous celebration of what appears to be simultaneously a house-inauguration and a marriage ceremony (Gaenszle 1993:202-203).

The culture hero, Khakculukpa, initially survived on the "fruits of the jungle." From firstly foraging, he progressed to agriculture and hunting. Agriculture and foraging are the two main subsistence systems used by the Rais today. However, the Kulunge Rai of Yangden do not now hunt as a subsistence strategy. They own no guns, nor bows and arrows. A few birds are caught for ritual purposes. They fish with baskets and fishing poles in the Apsuwa *Kholā*.

The origin myth helps to clarify the social context of how the farmers interact with their environment. The evil forest witch (*ban devī*) represents the many deities who reside in the *jaṅgal*, and Mansberger (1991:143) remarks that "she holds sway as protectress over all natural life." The Rais appease the deities through food offerings in order to prevent misfortune and ensure health for their family, livestock and village. The *jaṅgal* deities importance and relationship to present Rai culture is discussed in a later section of this chapter.

My Rai friends always said, "we are *jeṭha* (literally oldest son) in the origin myth." The following passage describes the early migration patterns of the Khambuhang, the ancestors of the Kulunge Rai, in Eastern Nepal. The Dudh Kosi is the area of Mahakulung where the Khambu Rais first settled (see Chapter 1, Figure 1).

> The – more or less direct – descendants of Khakculukpa are the four mythical brothers, Khambuhang, Mewahang, Limbuhang and Mece-Koce. Three of the brothers migrated from the mythical place of origin (*Khokwalung* M.) located down in the Tarai up towards the mountains to the north, whereas the youngest brother was left behind to stay in the

Tarai........ Eventually the brothers split up: Khambuhang follows the Dudh Kosi, Mewahang walks up the Arun Kosi and Limbuhang, who comes later, continues along the Tamur Kosi..... The descendants of the three brothers who had migrated along the three rivers in the course of time split up into different localities and settled down. This is the era of the village founders who established a special, transcendental link to a particular area, the *ca:ri*... meaning both specific, clearly delimitated territory which is claimed, as well as the divine "force" which is somehow inherent in it (Gaenszle 1993:204-205).

In the origin myth, the Kulunge Rai's relationship with their human brothers and the past are linked with communal rights to land and water resources. The Rais have a deep affinity with the *jangal* because the first human being's older siblings were plants and wild animals. Gaenszle (1993) shows that the narrative tradition provides a meaningful context for an understanding of the beings early origin which sets the stage for how *muddum* rituals actualize mythic events in the present.

The Environmental Messages of Myths and Legends

My intent is to look at the "hidden meaning" in myths and legends as important vehicles for environmental messages. According to Wright and Dirks (1983), environmental information that is essential to a culture must be encoded in symbolic form within narratives. Etkin (1994) asserts that these vehicles preserve a reservoir of knowledge about emergency plant species which may be otherwise lost. Seasonal food shortages are acknowledged in oral literature because there are numerous stories about dearth and hunger.

Legends are a method of non-formal education, often found in non-literate societies. Legends are prose narratives regarded as true, but not sacred, by the narrator and are believed to have happened in real time. On the other hand, myths exist outside of time. Human beings are the principal actors in legends, although their activities may take on a superhuman element because they may be aided or hindered in their adventures by gods or other supernatural beings. Traditional legends lend structure to issues of major importance for the community's welfare because they may have a moral slant validating cultural behavior and justifying the rituals practiced by the local people (Collins 1978).

Legends can function as a historical record of events of environmental adaptation or maladaptation. One young Sherpa man recited this legend which his grandfather had told him.

In the 15th Century, there was a big problem in Tibet because it was fighting with China. Many people left Tibet and headed south. They walked for six to eight days and did not see any villages. One day they arrived at Tongphuk.[26] *Barley was growing on the dry field and one yak was grazing in the area. They thought that this area was good and decided to settle there. They cut the barley for food. At first, they cut a little bit of the barley for a meal. The next day, it grew back again. However, they became greedy and cut all the barley. The barley disappeared and did not grow again. The people became hungry and because there was one yak, they said, "Lets kill the yak and eat it." They ate the yak, got a disease and all of the people died. Today the people's bones can be seen at Tongphuk.*

This legend can be verified by reference to known historic events. A group of Sherpas, who were escaping the turmoil of war caused by the Mongols in Kham, Eastern Tibet, arrived in Nepal between 1480 and 1500 (Oppitz 1968:75). This legend describes the Sherpa migration to a new environment (Nepal) which had abundant food supplies, but because of greed an environmental disequilibrium occurred.

This legend has both a moral and environmental message. The Mahāyāna Buddhist doctrine regards food as a necessary bodily fuel but something in which people should not over-indulge. "Gluttony, particularly, consuming meat or drinking blood, violated the Buddhist stricture against killing" (March 1987:379,f15). The killing of animals, eating meat and gluttony has had dire consequences for the population. The underlying message is transmitting a warning that a person should take only what is necessary for sustenance and preserve the environment because food supplies are not unlimited.

Diemberger (1992:CH4,33) stresses the importance of rituals and myths to legitimate political power and create "consensus as well as opposition in the relations between the small ethnic groups and the state." This point will be raised further in a · discussion in Chapter 9 of why traditional communal property rights are a current issue of conflict.

Description of the Ban Jhākri

The myth of the *ban jhākri*[27] reflects the social concerns of the local people about their environment. The *ban jhākri* is an intermediary between the forested-*jaṅgal* and the people. He possesses vast ecological knowledge which he derives from the forested-*jaṅgal*, his home. The Sherpa name for the *ban jhākri* is *buruni*. The

---

26 Tongphuk is the uppermost pasture area in the Apsuwa Valley used by the shepherds during summer grazing. *Tong* (T, *stong*) means 1,000 and *phuk* (T, *phug*) means cave.

27 Foreigners have described the *ban jhākri* as an autochthonous shaman (Maskarinic, personal communication, 1993), a nature spirit associated with natural phenomena, a specialist who helps [teaches] shamans and lives in the *jaṅgal* (Collins 1978), a *jhākri* of the jungle and a hobgoblin who inhabits mountain caves and entraps human beings (Turner (1931:419). Allen (1976b:536) classifies the *ban jhākri*, a tutelary deity or medium, within the blood-eating category because the *ban jhākri* is considered a potentially harmful spirit and demands blood sacrifices.

Kulunge Rai name for him is *rang keme*. According to my Rai and Sherpa informants, the *ban jhākri* is human-like male in appearance, has a small face, long brown hair, hairy arms and legs, and feet which are turned backwards. Based on this description, Parakram Yonzon sketched the *ban jhākri* (see Figure 6).

Figure 6. Sketch of the *ban jhākri* (by Parakram Yonzon).

The *ban jhākri* does not talk, but is considered to be human-like in that he can father children. The *ban jhākri* lives only in thick *jangal*, usually near a waterfall, steep cliff or stream. Once the forest is cleared, the *ban jhākri* leaves and looks for another thick *jangal* in which to live. The local people say that, "these days there are many people and less forest around here so we can not see the *ban jhākri*."

Paul (1976:145) points out that the symbolism of the forest and its local Nepali god (*ban jhākri*) is consistent with the idea of madness and wildness. During the shaman's initiatory craziness, he goes to the forest and behaves in a savage or uncivilized manner. Having established contact with the *ban jhākri*, his tutelary deity, the shaman can be introduced by him to the other supernaturals with whom he must deal. Among the Thulung Rais, a person can become possessed by a *buwalem*, rendered as a *ban jhākri* (Allen 1976c:134). The possession of a young boy by a *ban jhākri* is described by Turner (1931) in MacDonald (1976).

The *ban jhākri* can make people sick. This is the reason that villagers are afraid to walk alone at night in the *jangal*. Many taboos also exist to protect the unwary from receiving harm at the hands of the *ban jhākri*. If people throw a stone, spit or make a noise, the *ban jhākri* becomes angry and causes problems. It is because of this propensity to cause misfortune that the *ban jhākri* receives ritual propitiation from the Sherpas and Rais. I have seen picnickers make a food offering to the *ban jhākri* close to the picnic site. They say the *ban jhākri* will be happy and stay away from them.

The Legend and Setting of the Elusive Ban Jhākri

Until 1952, Nepal was closed to the outside world. In 1955, Nepal opened its borders to foreigners. There is a tale (in Collin's [1978] terms, a legend), dating from 1956 which tells of a foreigner who came to capture the *ban jhākri*. At this time, there were no planes and 20-25 days were required to walk from Kathmandu to the scene of the events. I have pieced together the story from the accounts of several elderly Rai and Sherpa informants living in Yangden and Gongtala. Literature published on Yeti expeditions in this area will help verify the truth of this tale.

In the year 2013 NS (1956), a strange bearded white man came to Gongtala. Before that, no villager had ever seen a foreigner. He was an American man named Peter.[28] Peter wanted to capture the *ban jhākri* and take it to a zoo in his country. He had a Sherpa friend named Dawa Temba Sherpa who accompanied him. They stayed about ten days in a tent at a sheep shelter by the waterfall near Gongtala. From a small hole in the fence surrounding his tent, Peter watched for the *ban jhākri* to appear. At that time it rained a lot and perhaps it was the month of *Jeṭh* (mid May-mid June). Peter called many people to the waterfall to help him catch the *ban jhākri*. He paid the Yangden Rais two Indian rupees for every frog caught because the *ban jhākri* likes to eat frogs. One sheepherder found the hair of the *ban jhākri* and showed it to the foreigner.[29] After seeing it, the foreigner was very happy. Each night Peter and the villagers with a torch looked for the *ban jhākri*. The *ban jhākri* got scared by the many people and disappeared. The villagers were afraid of the *ban jhākri* and returned to their homes. Peter was left alone. He saw the *ban jhākri* during the night in the black clouds.[30] But after it got dark, he could no longer see the *ban jhākri* and fired a gun near the waterfall. He was not successful in catching the *ban jhākri*. I think the *ban jhākri* got scared and ran away because the gun was fired and made a loud noise. After that the landslides came. The landslides occurred because the *ban jhākri* became angry."

---

28 In the spring of 1957, Peter Byrne and Tom Slick first explored the Chhoyang *Kholā* (Apsuwa) for three months. In 1958, Peter returned as part of the Slick-Johnson Nepal Snowman Expedition, with Gerald Russell, other members of the team and a Sherpa named Da Temba. "I chose upper Arun as it was one of the least known areas of Nepal, also because of stories I had heard of a 'tribe of hairy people' up there. I did not want to capture the Yeti, just find and photo/document it" (Peter Byrne, personal communication, 1994).

29 "Russell was given some reddish-brown hair by a local man who had found it in the cave..and may have belonged to a Yeti" (Dyhrenfurth 1959:324).

30 Da Temba with a local man saw in a stream around midnight a small Yeti no more than 4 feet 6 inches in height, possibly a young one. The Yeti had come for his nightly meal of frogs, and when a flashlight was aimed at his face he started toward the men. The next morning, Russell found small Yeti tracks (Dyhresfurth 1959:325).

In the story, landslides occurred when the foreigner tried to capture the *ban jhākri*. The villagers attributed this to the wrath of the *ban jhākri*. They believe the gods apply sanctions of their own against the wrongdoer which can be a major deterrent of deviant behavior. It is also true that landslides frequently occur during the monsoon season, and hail regularly destroys crops each spring so that a ceremony, *pūjā tos*, is often conducted to stop the hail.

Both Rais and Sherpas value the sacred property of the *jangal*. They have prohibitions that guard it because their survival depends on the best and most efficient way of using its available resources. They pay homage to the *jangal* deities in order to prevent the unpredictable and uncontrollable forces of nature from causing destruction. Thus the *ban jhākri* is a compelling vehicle for natural resource conservation.

A 350-year-old scalp and skeleton hand of a Yeti at Pangboche Gönpa (Mt. Everest) earlier suggested that two types of Yeti existed, a small species which lives in the dense Himalayan rain forests between 2,800-4,500 m, about four and a half feet high, and a larger animal, six to eight feet high, living in the higher regions between the uppermost villages and the glacier (Dyhrenfurth 1959:325-326). The smaller Yeti fits the description of the *ban jhākri* described earlier. McNeely et al (1973) disputes Dyhrenfurth's claim that there are two types of Yeti, because he feels the latter type cannot live in the snow fields (where the tracks were seen) where a food supply does not exist. The habitat of the smaller Yeti is described as impenetrable dense vegetation which would assure a sufficient food supply.

Sacred Realm: Indigenous Religion

Religious Healers and Practitioners

Syncretism occurs in these communities because some elements of Rai cosmology and rituals have been adopted by Sherpas while the Rais include the Buddhist concept of Beyul Khenbalung (Hidden Valley) in their "sacred interpretation of landscape" (Diemberger 1992:CH3,38). Maskarinec (1995) has pointed out that shamans are active in all life cycle activities among all ethnic groups in Nepal who follow Hindu and/ or Buddhist beliefs. The types of gods and deities and spirits worshiped by Sherpas and Rais differ. The Sherpas focus on mountain gods while the Rais, who formerly hunted, center on hunting spirits and deities of the *jangal*. However, both groups have a tradition of *jangal* plant use for ritual and symbolic purposes.

Many of the same rituals are performed by Rais and Sherpas because they share similar beliefs about the causes of illness and the healing treatment. Public rites;

seasonal rites; domestic rites, such as births, name giving and funerals; proprietary rites; and curative rites for humans and livestock, are observed by one or both groups in my research area.

In rural and remote regions throughout Nepal, traditional ritual practitioners play an important role in the welfare of the individual, household and community. Home remedies which include indigenous medicinal plants (*ja ibuti*) and food prescriptions or proscriptions may be the first course of action taken by the villager with the hope that the disease will be self-limiting. If the illness continues, the villager consults a traditional religious practitioner, that is, a shaman (*dhāmī*), priest (*pūjāri*) or *lama*. Local folk or traditional health care is more accessible and the healer offers a satisfying and culturally meaningful interpretation of the illness.

Both *dhāmīs* and *lamas* were interviewed in my research area using a semi-structured format which looked at the major rituals performed and their purpose, the flora and fauna used, and the types of illnesses they treat. Traditional health practitioners treat patients with symptoms of: fever, headache, stomach ache, fast heart beat and dizziness. They refer patients whose illness is caused by 'germs" and not by a deity or spirit to Western facilities, the local health center or the hospital in Khandbari. The symptoms which warrant referral are: measles, diarrhea and coughing up blood. The British Nepal Medical Trust set up a training course for shamans in Khandbari, to assist in the identification and referral of individuals stricken with tuberculosis.

The Kulunge Rai

The Kulunge Rais practice an indigenous religion minimally influenced by Hinduism and Buddhism. The Rai religion is based on a body of myths and ritual recitations aimed at the periodic appeasement of ancestral deities and spirits associated with nature (Gaenszle 1993). The Rais have personal and ancestral relationships with their land which have been defined by their origin myth, and also embodies their vast indigenous knowledge of the *jaṅgal* and its resources. The *jaṅgal* is viewed as an abode for deities and spirits living in the forests, mountains, rivers and lakes. Wild animal spirits are expressed in the rituals. Respect is shown towards these deities and spirits by the farmers through *pūjā* rituals and food offerings. Every Rai household also has a household deity. These rituals are believed to help maintain harmony and prevent misfortunes caused by the god's displeasure. One Rai said, landslides occur because the *bhūme* which belong to the earth, and *devī* deities, who are ancestral deities linked with the territory, have become angry.

Shamans (*Dhāmīs*) and Priests (*Pūjāris*)

The terms used to refer to shamans differ throughout Nepal and are locale specific. I use the villagers terms and contextual meaning for these religious practitioners. When I began my research in Yangden, there were three shamans referred to as †*hūlo dhāmīs* (large) by the Rais. There were eight priests (*pūjāris*), referred to as *sano dhāmīs* (small). Both shamans and priests perform religious functions of a general nature, and rites to ensure the health for all household members, treating illnesses and healing the sick. The shamans differ from priests in that through possession they act as mediators between humans, deities and ancestral beings. Aided by a tutelary spirit, shamans can divine the future and have the power to communicate with the deities residing in the *jaṅgal*. Their main responsibility is to conduct the major rites for the village.

The shaman Bijale was considered "the greatest *dhāmī* of all." He was killed suddenly in a horrifying accident just prior to completion of my research. When Bijale was repairing the stone ledge wall which bordered his agricultural fields, a huge stone fell on top of him. He was pinned to the ground and only his head showed. It took five men to roll the huge stone off him. He passed away shortly after this accident. While he was alive he was often summoned to the neighboring villages of Chhoyeng, Waleng, lower Walung, Yaphu and Seduwa to perform various religious rites because, as everyone said, "there was no one who could match his power in those places." After he passed away, the general consensus was "there will be many deaths because Bijale has died." This belief refers to the fact that he was no longer available to heal people. Human souls who have died violent, unnatural or inauspicious deaths may give rise to illness-causing wandering spirits (*bāyu*), and food offerings are often given to appease them. Priests conduct *pūjās* for patients who have psychological and psychosomatic complaints. They treat people attacked by witches (*bokshoi*), ailments of unexplained phenomena "due to supernatural cause" or bad spirits (*bhut*).

The Sherpa *Lamas*

The Sherpas are followers of Nyingmapa, the oldest sect of Tibetan Buddhism, founded by Guru Rinpoche (Padmasambhava) in the 8th Century A.D. In all the Sherpa villages, *lamas* (Buddhist priests) help villagers achieve religious merit. The *lamas* became religious practitioners because they wanted to follow the tradition of their forefathers who were also *lamas*. Their main function is to conduct the *Daśamī* ritual, a monthly feast and celebration in honor of Guru Rinpoche because Khenbalung is his "resting place." *Daśamī* (T, *tshes.ba bcu*), occurs ten days after the new moon. *Thulzhi* (T, *aphrul zhig*) Rinpoche, the head of the Nyingmapa sect at the Thupten Chöling *Gönpa* (in the Solu region), and other great Rinpoches are honored at this ritual. *Torma*, (T, *gtor.ma*) ritual offering cakes, usually made of barley flour and decorated with butter, represent the various gods. These *torma* are given as food offerings to the gods and shared with the attendees at the end of the ceremony.

Sherpas seek out *lamas* when they are sick. The *lamas* recite mantras, read the Nyingmapa holy texts, light butter lamps and use *tu* (T, *akhrus* [holy water]) to purify the patients who have been polluted. An offering of *chang*, butter, rice and ginger may be made. The *lamas* also uses medicinal plants: *chiraito* (*Swertia chirata*) to treat colds; *siltimur* (*Litsea citrata*) and *namcharok* (S) (*Phyllanthus emblica* L.) to treat diarrhea; and *timur* (*Zanthoxylum armatum DC*) to treat stomach pain.

The lamaistic and shamanistic systems coexist in Sherpa society. Shamans are sought out by Sherpas for diseases that have purposeful supernatural agents (Paul 1976). "According to Sherpa belief, disease itself marks spiritual impurity" (March 1987:380,f19). In Gongtala, I saw both a Rai priest (*pūjāri*) and a *lama* treating a patient who had been stricken by a ghost (*bhut*). If Sherpas living in Saisima want a shaman they must go to Dobatak because animals are not killed in Saisima. In the pre-Buddhist tradition, animals were sacrificed because they were linked to mountain gods. Animal sacrifices were prohibited after Buddhism was introduced (Diemberger 1992).

Religious and Healing Rites

The present rituals of the Kulunge Rai and Sherpas relate to the origin myths, *jaṅgal* deities and plant use. I will begin by discussing the main religious and healing rites practiced in Yangden and the Sherpa villages by the religious healers.

The *Nāgi* (*Lu*)

The *nāgi* (a snake and the original creator of mankind) is considered a universal deity. The shaman performs the *nāgi* rite for either a household or an entire village. Various rites are performed and sacrifices are regularly made to this deity as ancestor worship (Gaenszle 1993; McDougal 1979).

One communal rite to the *nāgi*, referred to as *bhūme* (earth or agricultural rite), is performed by the shaman just before the millet and buckwheat crops are to be harvested. The rite ensures that the first fruits of the crop harvest will be bountiful. The supernatural "agencies which control the forces of nature" are placated during this rite (McDougal 1979). In 1992 in Yangden, twenty chickens were communally sacrificed for the ancestral deity at the *nāgi*. *Jāḍ* (Rai maize beer) served in gourds is offered to the ancestors.

Chang phi (T, *chang phud*) is an annual celebration held by the Sherpas before the herds depart for the high pastures at the *lekh* during early summer. It honors the *lu* (T, *klu*) of pre-Buddhist origin, water spirits which dwell around lakes, rivers and small streams. The *lu* are identified with the *nāga*. The *lu*, normally benevolent spirits, can become angry if their dwelling place has been polluted and will cause illness (Cornu 1990 in Samuel 1993; Wangmo 1990). One foreigner who was staying at Saisima had washed her hair in the spring nearby the village and "polluted it." The Sherpas told her they had to do a cleansing ritual and later they were seen placing incense and ashes on a rock over the spring and reciting mantras to appease the *lu* deity. The people said if this ritual was not performed, the *lu* would become angry and make the people ill. The *chang phi* ritual is named for the barley drink which is offered to the *lu* since Sherpas, as Buddhists, do not sacrifice animals. Only the first water poured off of the cooked mash is used because it is considered very sacred. The *chang phi* is offered to the ancestors to ensure the well-being for the household and their livestock. Downs (1980) explains that a belief in the *nāgas* [*lu*] forces a respect for.... any water source because, if they are polluted, the *nāgas* which live at the source will retaliate. This also encourages good sanitary practices, as it is forbidden for humans and dogs to defecate or urinate near the spring. This does not hold true for livestock.

Another group of sacred beings are the forest spirits, territorial beings which represent the 'wildness' of the *jaṅgal*, inclusive of the rivers, lakes and mountains. They control the forces of nature. They are dangerous, and offerings must be made to appease them. Many households told stories of drownings in the Apsuwa *Kholā*. One Rai shaman said that if a person is scared when crossing the Apsuwa *Kholā*, he or she will become sick. A *pūjā* must be made and a chicken sacrificed for the river spirit. In Gongtala, before timber is cut for a new home, a *lama* performs a ritual. The people are afraid that if this ritual is not done, they will anger the *jaṅgal* deities who will cause misfortune. The Sherpas consecrate the main parts of the house at various stages of construction and bless it with butter and barley flour.

Pigg (1989) says certain spirits are linked to certain cultural groups. I also found that many of the deities worshipped in the *pūjās* have specific ethnic identities, for example, *Mulu śikāri* (a deceased Rai), *Goṭhālo jhākri* (spiritual soul of a Tamang) and *Burheni* (incarnation of Kung), while some are named after places, *Yaphu Devī* (incarnation of Yaphu). *Śikāri* is always included as one of the deities for which a *pūjā* is held by the Rais. A Yangden Rai shaman explained why the *ban śikāri* deity, who represents the spirit of the hunt, is of particular importance to them.

> *The name of our deity is the ban śikāri. Ban śikāri means hunter of the forest. A long time ago the Rais were jaṅgali people.*[31] *Their main compassion was hunting and eating wild animals, such as deer and bear. One day a Yangden Rai hunter went to the jaṅgal to hunt, but died there. His soul still craved meat and he was transformed into a śikāri deity. Today the śikāri has become the spiritual soul of the Kulunge Rai. Śikāri can make the Rais sick so this is the reason why today we must worship them.*

When I accompanied six Rai shepherds to the high mountain summer pastures in the Himalaya, I noticed that once we reached a certain point in the transhumance

---

31 R.L.Turner (1980:206) describes *jaṅgali* as, "wild, savage, uncivilized, rustic, a wild man."

move, a daily *pūjā* ritual was conducted. The offering plates are made of *pāteliṅgā* leaves held together by a anthropogon sliver of bamboo. A *chimal* flower with dried *sunpati* (*Rhododendron anthropogon*), sheep milk, butter, rice, ginger, fish, bird feet and chili are included in the offering. The *pūjā* is held daily during the entire three-month summer of high-altitude grazing to appease the deities and ensure protection for themselves and their livestock. The shepherds fear that the deity will become angry and harm them if they abuse the environment.

Leaf plate offerings at the *lekh*

"In a Buddhist perspective, mountain deities are protectors of religion" and mountains serve as bridges between the world of the people and that of the gods (Diemberger 1992:CH4,33). At all high passes, *khatas* (T, *kha.btags*, white prayer flags) are seen blowing prayers to the Gods into the wind and asking protection for the journey ahead. When Sherpas are cremated in the forest, *khatas* are put up. At the *lekh* during the summer months, the Sherpas set a small amount of rice and milk pudding (*kir*) on leaves to worship the mountain gods and ensure health for their animals and their own welfare as well.

Deva and Shamanistic Trance (*Cintā*)

During the full moon in *Dasaĩ*, I observed two rituals, *deva* and *cintā*, practiced by the Kulunge Rai. The *deva* ritual has to be celebrated twice a year in spring and autumn in combination with a shamanic seance (*cintā*) that lasts through the night. *Devas* are ancestral (lineage) deities linked with a specific territory. They are responsible for the safety and prosperity of the household and the village (Gaenszle et al 1990). This ritual began with a sacrifice of a pig in the afternoon at Bijale's house and continued throughout the night until the next morning. The pig sacrifice appeases the deities and thereby diverts disease and evil from the people. The consecrated gifts (*prasād*) were distributed among participants as blessings from the ancestors. The ritual protects against landslides, hail stones and windy storms.

The Sherpas also conducted a *deva pūjā*, but without the *cintā* seance. In Gongtala, the *deva pūjā* is performed by a shaman twice a year and one member from each household participates. Each family contributes money which is used to pay for the sacrificial animals (goat and chickens).

The *cintā* is held annually in conjunction with the *deva pūjā,* either as a prophylaxis, or in response to a particular medical problem "when simple day-time exorcism has proven ineffective" (Allen 1976b:31). One Rai shaman explained that the *cintā* is conducted because there is neither medicine nor are there doctors in the village. The ritual cleanses the house and acts as a precaution against the anger of the ancestral deities. If one member of a family does a *pūjā*, then it will include all extended family members. During the *cintā* ritual, Bijale wore elaborate attire which included a feather headdress of wild pheasant feathers and antlers which are "symbols of the hunter" (Gaenszle et al 1990). The wild bird species are symbolic of celestial flight (R. Jones 1976:34-35). Stalks of 'wild' bamboo (*phurke*) were waved during the seance. Allen (1976c) explained that it is used by the Thulung Rai shaman "to extract the evil from the patient's body."

The *lamas* compared their *domang* (T, *mdo mang*) ritual to the Rai *cintā*. A senior *lama* reads various sections from the holy texts *pecha (T, dpe.cha)* to purify the household and ensure that no one will get sick. The *domang* can take up to three days to complete because of the length of the texts. A Sherpa family in Gongtala whose baby died after falling in the fire had sent for the senior *lama* from Saisima to perform the *domang*. After the *domang* was completed, the *lama* said that no more misfortune would strike the family.

Sansāri aitabar are universal divinities responsible for epidemics (Sagant 1976:93). The Kulunge Rais said they follow the practices of the ancestors and hold a *pūjā* for *Sansāri aitabar*. The ritual is sometimes done once yearly to prevent the deities from becoming angry. They believe if a *pūjā* is not held, they will become sick, have heart problems, become dizzy, and suffer with joint pain. Every year after this *pūjā*, the shaman said the people get well. The entire village of Yangden does it to protect itself from major diseases.

Bijale performing the *cintā*

Healing Ritual: A Personal Account

I accompanied six Rai shepherds and their hundred sheep to the high mountain sheep shelter during the monsoon because I wanted to observe how they use and manage the resources of the *jaṅgal* and *lekh*. In the beginning of the journey around noon one day, I slipped and fell on the muddy trail and sprained my wrist. I cried from the pain, but ran to keep up with the pace of the sheep. During the trip, my feet became blistered. After several days, when we reached the highest pasture area at the *lekh*, about 6,400 m, my feet had become infected. We rested one day before making the descent back to the village. The Rai shepherds said, "We must return within two days to our village to help with the harvest." I put my swollen feet into my hiking boots and laced them up for the descent.

When I reached Gongtala, I literally collapsed. My feet were infected and swollen, and I had a fever and diarrhea. All my first-aid medicine had been used on the villagers and there was none left. Because I was anorexic, my village friends realized I was ill and became very concerned for my welfare. (Messer [1989] has described the attention shown towards eating behavior marks a culturally-recognized illness). One Sherpa *lama* brought me religious butter lamps. Everyone suggested a *dhāmī* look at my feet. He first checked my pulse and then performed a rice divination. Both my pulse and the rice divination helped him determine the cause of my illness, the place it occurred, and the course of action to be taken (R. Jones 1976). The shaman said that I had been attacked by two *jaṅgal* deities (*śikāri* and *Burheni*) who had entered my body when I sprained my wrist at the *lekh*. The shaman said I was frightened and had cried out, which attracted the deities which possessed me.[32] The shaman then used a procedure referred to as *jhar-phuk* (sweeping and blowing) in order to remove the deities and cure me of the affliction by expelling the deities from my body. After the exorcism he said, "You will get well and be able to walk out to Khandbari" (a two-day walk). With the additional help of gentian violet from the Tamkhu Village Health Post, I was able to make the trip.

Offerings and Sacrifices Used During Rituals

The *pūjās* held by shamans use both 'wild' and domestic plant and animals. The cock is said to represent the spirit of the hunt. The life force of the animals is sacrificed for the life force of the living. Chickens are frequently required for the taking of omens, which are based on the scrutiny of the livers. The chickens are important, not only for their meat, but also their eggs. The egg is more auspicious than the chicken because a shaman can tell the future from the quality of the yolk (Gaenszle et al 1990). A huge banana leaf is placed in the fire, and an egg is broken into it during the *pūjā*. The viscera (liver and other entails), part of the food offering, are examined numerous ways by different shamans (Maskarinec 1995), and the heart and liver are offered to the *śikāri*.

Usually one cock or hen is sacrificed for each god. In practice, the number of animal sacrifices depends on the household's economic status and which animal is available for the sacrifice. A poorer household may sacrifice only baby chickens. One Rai shaman said the deities have preferences for particular types of animals used for the sacrifice. *Singha Bahini devī*[33] needs a goat, *Bhūme* needs a cock, *Burheni* needs a hen and *śikāri* needs a cock. A sacrifice of a chicken, goat or duck must be made to

---

32 The deities are out at ambiguous points of time, that is, dawn (when night changes into morning), noon (neither morning nor afternoon) and dusk (beginning of darkness). Spirits attack those whose vitality has been lowered, that is, the normal harmony within the body is disrupted (a shaman at the immigration office in Kathmandu, personal communication).

33 *Singha Bahini* is a lion goddess that is believed to have clawed her way out of the earth (R.Jones 1976:40).

the *ban jhākri* deity to prevent people and animals from becoming sick. The fish, flesh of the *pāhā* (Garden toad) and legs of a pheasant are also needed.

A sacrifice performed by a *dhāmī*.

The religious plants used in *pūjās* are: *shukpa* (S) (*Juniperus incurva*); *asari* (*Mussaenda frondosa*); *sunpati* (*Rhododendron anthropogon*), and *khenba* (S) (*Artemisia vulgaris*). Also, *dona*, *ambang shermak* (S) and *katush* are used. Among these plants, 'wild' lavender, referred to as *titepati* (N) *or khenba* (S) (*Artemisia vulgaris*) seems to have universal religious importance for Sherpas, Rais and other groups. It is used during the *deva* ritual by both Sherpas and Rais. Beyul Khenbalung, the legendary "Hidden Valley," is literally the "hidden valley of the Artemisia" where different types of *khenba* grow (Diemberger 1992). Several anthropologists who described various shaman rituals mention 'wild' lavender. For example, Stone (1983:974) says the shaman used a weed called *titepāti* that gives off a bad odor and is said to attract evil spirits away from a Brāhman woman who had been attacked by a demon. 'Wild' lavender has mystical cleansing properties and is also used by Limbus to ritually cleanse themselves after a funeral (S. Jones 1976:24).

A purification ritual was performed for a man sick with fever by his father who was a tribal priest. The leaves of the 'wild' lavender were used to sprinkle water during this ritual (Sagant 1976). During Rai weddings, water was sprinkled with the 'wild' lavender when the bridegroom arrived. It is also important in making *chang* and is discussed below in the alcohol section.

Aromatic plants used by the *lamas* as incense are: *shukpa, khenba, purmang* (S), *pom* (S) (*Juniperus wallichiana*); *takpa* (S) (*Betula utilis*) and *masur*.

The foods offered to the deities usually consist of butter, ginger, rice, egg, all of which are considered prestigious foods because they are rare and often must be purchased. 'Wild' ingestibles are considered to be unworthy offerings for deities because they come from their own domain. The farmers only give foods which they grow, livestock (and their products) which they raise or food items that are purchased.

Titepāti plant at a Rai wedding

Other Categories

Maghe Sankrānti

The symbolic significance of food from the *jaṅgal* for the Rais is that it ties them to their native land. Their reasons for settling in Yangden was the abundance of *ban tarul* (*Dioscorea versicolor*) and *ghar tarul* (*Dioscorea alata L.*). These foods are included in a Hindu celebration called *Maghe Sankrānti*. The ritual occurs at the beginning of the month of *Maghe* in the Nepalese calendar, which marks the agricultural new year. The Hindu celebration is celebrated by all Rai households. *Ban tarul* and *ghar tarul*, chayote, rice or millet unleavened bread cooked in oil, chicken and pork are eaten and maize *raksī* is drunk during this celebration. A *pūjā* is not held.

Saune Sankrānti

Saune Sankrānti is another celebration observed by the Rais, Limbus, as well as Hindus. The symbolic significance of this festival was described by a Limbu.

> *This festival is celebrated in order to send away all the diseases and animal hunger and help welcome a healthy environment and sahakāl (prosperity). In our village, people tie all kinds of vegetables in a rope and hang them around the house. At night, a nānlo (a bamboo winnowing tray) is beaten with a stick making noise and the person says, "Go away diseases and anikāl and let sahakāl come to us." The nānlo is taken to all the corners of the house and finally it is thrown away. We believe that along with the nānlo, all kinds of illnesses and anikāl will go away. On this day, we celebrate this festival by eating meat and other good food* (Dovan Lawoti, personal communication, 1997)

These festivals, *Maghe Sankrānti* (mid-January) and *Saune Sankrānti* (mid-July) are held to divide the year. *Sankrānti* marks the agricultural season. This festival coincides with *anikāl* (food shortages) and the 'wild' tubers are major staples during this period. Besides their nutritional value, they also have symbolic importance.

Poisons and Plants

"Medicinal plants are viewed as cultural objects and as biodynamic elements that have infinite pharmacologic potential" (Etkin 1993:93). *Bikh (Aconitum)* is a poisonous plant which exemplifies the importance of culture and context in the assessment of plant medicines. *Bikh* is exported to India and used in medicinal preparations as discussed in Chapter 5. In regions where it grows at the *lekh*, there are many stories of deliberate poisonings and it is a common topic of conversation around the hearth in the evening.

The Rai shaman said there are three types of *bikh:*[34] the venom of the snake, *bhyākuto* (frog) or *pāhā* (garden toad) and poisonous plants which grow at the high pasture referred to as *lekh bikh*. These will be the focus of this section. *Lekh bikh* can not be put into hot things, for example, tea. It is placed in cold alcoholic beverages. People are cautioned not to drink these beverages in stranger's homes. Certain villages are known for poisoning. This is also true among the Khasis of Meghalaya (Standal, personal communication, 1997) and in Kongpo, Tibet, where *aconite* grows (Daniggelis 1996).

---

34 *Bikh* is a general word for poison (R.Turner 1980:439).

Only women are said to give *bikh* because men tend to fight and relieve their anger, while women tend to repress anger. The symptoms of *bikh* poisoning are: bodily aches, painful armpits, fever and difficulty breathing. The shaman does a rice divination and indicates if the person has been poisoned. After the patient eats the antidote, *bikuma*,[35] watery diarrhea and vomiting occurs, the *bikh* leaves the body, and the patient is said to get well.

There are many stories of poisoning in Eastern Nepal and in Kongpo, Tibet, particularly in areas where *bikh* grows. All the villagers talked about a poisoning which occurred in the village of Walung, a two-day walk northeast of Gongtala, in February 1993.

During the month of Phalgun, ten people were poisoned at a Sherpa wedding in Walung. Prior to the wedding, several of the women had argued with the groom's parents and 'held a grudge.' They became angry and there was an argument. They said, "At your son's wedding we will do harm to you and give bikh." The parents felt this wouldn't happen. During Sherpa weddings, it is a traditional practice for women to bring their own homemade chang to the wedding. Most of the wedding guests had left the groom's home and gone ahead to the bride's house. Ten people who came late drank the chang, which had bikh. Four people, three Sherpas (including a pregnant woman) and a Kulunge Rai died. The other six people who survived had eaten dog feces, which acted as an emetic by inducing vomiting which brought up the bikh. Eight policemen were summoned from the Khandbari District Police Headquarters. The women who had poisoned the guests hid in the forest, but they returned to the village because they were hungry. At this time, the women were seen by the villagers and the police were notified. Three women were taken by the police and put in the Khandbari jail. They were sentenced to twenty years.

I have been interested in the notion that only women are said to use *bikh*. Ortner (1995:386:28) said poisoning is thought to be done by women as a means to enrich their households. When the poisoned person dies, the poisoner "somehow gains his or her wealth." Diemberger (1992:CH7,18 and 33) notes that "since all things inherently contain both positive and negative powers, women, who are the protagonists of kin-tie spinning and food distribution (positive), are inevitably also held responsible for the negative power of food, which is poisons (in conflictual cases men are considered to use black magic rather than poison). The ritual represents, in any case, also a clear expression of religious control over the ("negative") power of women."

---

35 The plant *bikuma*, also referred to as *semen* (S) or *phongma* (T. *phong.dmar*) is considered the antidote for *bikh* and is also found at the *lekh*. *Bikuma* is luminous in the dark. It grows in the same place as *bikh* and can only be picked at night in the moonlight. The four types of *bikuma* are distinguished by their color: white, black, red and yellow. *Bikuma* is bought in the market by villagers and carried when they travel to other places where it is said that *bikh* is given. *Bikuma* is pulverized with water and eaten in a raw form, which has a bitter taste. It is used to treat stomach ailments and fever.

Cultural Views of *Jangal* Ingestibles

The situational level of a person's ritual purity varies continuously during the life of every individual due to involuntary factors (Ferro-Luzzi 1980). Food consumption is culture-bound and food proscriptions emerge during rites of passage, such as stages of child-bearing (pregnancy and in the puerperium and lactation period); among certain social groups and for ritual purposes. The symbolic attachment of 'hot' and 'cold' to certain foods, including *jangal* ingestibles, prevails throughout Nepal, as well as the rest of Asia. According to Simoons (1991), it is unclear whether the hot-cold classification system has a single origin (places as varied as India, Arabia, China, Greece, Central America (Maya) or Southeast Asia have been suggested) or had independent origins in more than one region.

Few food avoidances at puberty are observed in either Rai or Sherpa culture. Elderly men and women (beyond menopause) observe few, if any, food taboos (Simoons 1994). An elderly Sherpa informant said, "Cow's tongue is only eaten by the elderly because it can cause a person to have slurred and difficult speech. Brains are also only eaten by the elderly. If brains are eaten by younger people, it is believed they would get dizzy and not be able to cross bridges." Brains are cooked separately and said to have a 'bitter' taste, a quality often associated with cold foods.

The traditional medicine systems of India and China may have influenced the present hot and cold classification used by the people in my research area:

- The Indian diet, influenced by Ayurvedic and Unani, centuries-old systems of medicine, classifies foods as "hot" or "cold," according to the effects they are believed to have on the body under various conditions and in different seasons (Golpaldas et al 1983b).
- The 'hot' and 'cold' classification of the Chinese is based on metaphorical associations among people, plants and the universe. These hot and cold food categories are rooted in China's ancient Yin-Yang philosophy of conflicting yet complimentary opposites. "Cold" (yin) food, must be in balance with "hot" (yang) food, which is said to create harmony, enhance health and prevent illness (Wu 1979). Their heating and cooling effects on the body dictate how and when they are consumed or avoided (Lindenbaum 1977). Cold foods are proscribed because they are associated with illness and disease. According to this classification, curcurbits, leafy vegetables, fungus and most fruits are "cold" foods, while meat, eggs, carp, spices (black pepper and chili) and alcohol are "hot" foods. Cold foods have a higher percentage of water, less protein, fat and carbohydrates and fewer calories than hot food (Manderson 1981). In Chinese culture, gender is (classified) as hot or cold, men are said to be yang and women, yin. Women are

naturally inclined toward the cold side and are considered at special risk when their femininity is most pronounced (Wu 1979:24) which coincides with nutritionally vulnerable periods, for example, menstruation, early pregnancy and lactation. Menstruating women lose blood, a hot humor, and therefore must eat hot food to restore bodily balance (Simoons 1991; Wu 1979).

Childbirth

In both Rai and Sherpa culture, childbirth is ritually "polluting" to the mother, infant and all household members where the birth takes place. This is in agreement with the findings from Levitt's (1988) study of traditional birth attendants and perinatal development. The state of childbirth is medically, culturally and ritually significant. Sherpas view birth as perhaps the most concentrated source of pollution. Because birth is created by sexual intercourse, the newborn infant itself is an excretion, and is born dirty emerging "amidst blood and gore." The birth is felt to begin another cycle of biological growth and decay. The *lama* performs the *Tu* (T, *akrus*) ceremony (which cleanses the baby and family with holy water) and purifies the pollution which normalizes the balance between the physical and spiritual (Ortner 1973).

Childbirth for a Rai or Sherpa woman usually takes place at the home of her affinal relations, because residence is often patrilocal. The mother or a female relative or friend cuts the umbilical cord. Usually the husband helps with the preparations prior to delivery. After childbirth, the mother and her baby, including the area where the birth took place, is considered ritually polluted. For three days (Sherpas) or five days (Rais), the mother must stay in a separate area of the house in order to observe "pollution rules." Only immediate family members are permitted to enter the house. The area where the mother and infant are confined is located away from the hearth (sacred area). Male family members are not allowed to approach the mother during this period. Only her mother-in-law or another female can go near her. All of her food must be cooked on a separate *culo* fireplace by her mother-in-law because she is not allowed to cook for herself or others.

Prior to the name-giving ceremony, the raising of the "head-soul" ritual is performed. Gaenszle (1996:77-93) describes "raising the head-soul" (referred to as *saya po:kma* by Mewahang Rai), an important rite celebrated after birth by the Kirantis of Eastern Nepal. The mother's *saya* is said "to fall" at the time of pregnancy, due to the difficult birthing process and the first few days puerperium. According to Rai tradition, "her *saya* must be raised." The author emphasizes that it is metaphysics of vital forces forming the base of the rite which requires periodic renewal, and not pollution.

A chicken is "sacrificed" according to the Rais. The chicken is not simply a present to the ancestors but is regarded as an ancestor itself. In the origin myth the chicken is the first born and the eldest of the living creatures. The chicken went ahead of all the other animals on the route of migration and showed human beings the way. Because the chicken is more senior than man in the origin myth, it is said to be able to raise the *saya* (head soul). Incense (*dupi, Juniperus incurva*) is burned and wild lavender leaves (*Artemisia vulgaris*) are used for purification during the ritual.

Ritual Fasting

Ritual fasting, as practiced by *lamas* and monks, is usually done as part of meditation. They abstain from certain cultivated as well as *jangal* foods for health reasons. For example, when the head *lama* at Saisima was in a two-month meditation retreat, as part of his meditation practices, he avoided leafy greens: *sisnu*, rape leaves (*Brassica napus* L.); cultivated tubers: radishes, potatoes, and 'wild' *jangal* tubers. The *lāmā* said, "these foods contain poison which can burn my tongue." The "poison" may refer to the bitter taste caused by toxins which are a common property of *jangal* foods.

Status and Cultural Significance of Various Foods

Prestigious Foods

Food has social, cultural and psychological significance which can affect an individual's well-being. The motivation underlying food choices is a very complex issue based on local ideologies which vary with the individual circumstances. Food is a very powerful symbol of the upper class and more affluent groups and foods associated with them are considered prestigious. The prestige of these foods is not based on individual preference but on expense and status. They also may not have the highest nutritional value. Prestigious foods often must be purchased, are expensive and therefore beyond the means of the poor. They are used in rituals, served or taken as gifts to other households during special religious celebrations, such as, *Dasaī* , *Losar* (Tibetan New Year) and weddings. These foods include: livestock and livestock products: cream, milk, butter, buttermilk, meat and eggs. The ownership of livestock is therefore a symbol of wealth.

The pig is the sacrificial animal for weddings and *Dasaī* , and other rituals, and pork is the most highly prized meat. Both Sherpas and Rais like the fat of meat, especially pork. At a Rai wedding, accompanying the *jād* and *raksī*, a piece of meat is wrapped in a leaf for each guest, usually pork or mutton, an indicator of its prestige. At the wedding, after money is given to the bridegroom, the guest receives *raksī* and *jād* and a piece of pork fat. Fowl and eggs are also relatively expensive. This is the reason that eggs (fried in a lot of oil) are often mixed in *raksī* or *chang* and served to distinguished guests by Sherpas.

Rice is a high status food because it must be purchased and is expensive. Ortner (1978:14) claims that rice "is essential to the ideal production of hospitality." As a luxury food, it is served at Sherpa weddings and other important celebrations and rituals offered to deities during *pūjās*. Although rice is always served at Sherpa weddings, the Rais serve boiled coarse maize (*chakla*) to wedding guests. At a Rai wedding I attended, a small amount of rice was served with the *chakla* because rice is very prestigious, and is rarely eaten by the Rais.

Sanskritization

Sanskritization, a term introduced by Srinivas (1956), describes a process by which a caste undergoes cultural transformation with the hope of improving their social status so as to raise themselves to a higher position in the Hindu caste hierarchy. Berreman (1993:386) describes the implicit meaning of the term "*jaṅgali*" and the implications for the group who are associated with it. Rais have often been referred to as "*jaṅgali*" (literally of the jungle), meaning backward and uncivilized, by the high-caste Brāhmans and Chhetris. Even the Sherpas refer to the Rais as "*jaṅgali*." When Tibetans first settled in the area of Beyul Khenbalung inhabited by Tibeto-Burman peoples (Rais), the Rais were viewed as 'wild,' having no monasteries nor proper writing (Diemberger 1992:CH 4,2). Berreman (1993:386) further asserts that the term "tribal as it is used in India refers to people whose culture does not historically, or at present, include certain economic features such as plow agriculture nor do they follow one of the great religious traditions (Buddhism, Hinduism)." They follow indigenous religions and live in peripheral areas, such as mountains. The term tribal has the implicit meaning of unfamiliar rituals, diet or scanty clothing. In the Sanskritization process, people begin to avoid foods associated with the *jaṅgal*. Food is a symbolic marker of caste status, and low status foods are associated with the tribals. *Jaṅgal* ingestibles ('wild' plants and mushrooms) are viewed as easily accessible and "free" and thus their importance has been devalued. In these communities, 'wild' foods, foods gathered from the *jaṅgal* are stigmatized as 'poor person's' food and 'famine food.'

During the research, evidence appeared of changes in food preferences which are occurring among economically better off Sherpas who are striving to attain a higher status through becoming modern and richer and who are also undergoing a type of Sanskritization. These wealthy Sherpas showed a disdain for *jaṅgal* edibles when guests were eating at the family hearth. However, although these foods are considered a "poor person's food," they still said they like the taste, and the increased variety in their diet.

Sisnu carry a social stigma among Sherpas who are economically well-off. These Sherpas apologized whenever *sisnu* was served to me, even though I relished the taste and variety this added to my diet. The other poorer farmers were delighted whenever I ate *jaṅgal* ingestibles in their homes, especially when I would readily accept a second helping. A male foreigner who was conducting research in a Sherpa village, a two-day walk to the west, had never been served *sisnu* in the entire year he stayed there. My Sherpa friend who was poor and had accompanied me to his village boasted that I loved *sisnu* and other *jaṅgal* foods.

The status of *sisnu* in the eyes of Sherpas can be seen in the behavior of a Sherpa who hired several Rais from Yangden to help build a hotel. He paid them substantially less than the Sherpa laborers, worked them harder carrying rocks and fed them badly. Their meals consisted of a thin porridge and *sisnu*, suitable in his mind only for Rais.

In the Gorkha District, a United Mission/Nepal primary school teacher said that some youngsters complained of night blindness. In this case, interestingly, it was the higher status children who suffered because they lacked Vitamin A while the low caste children who ate *sisnu* experienced no eye problems (Krantz, personal communication, 1991). These nutrient consequences of Sanskritization have been extensively studied among tribals in Gujarat by Golpaldas et al (1983a and b). Inadequate intakes of Vitamin A, iron and ascorbic acid were found among the Sanskritized groups which had foregone tribal customs of eating meat, alcohol and "forest" foods. Tribal children were found to have higher weight-for-age values than non-tribal children. The tribal groups, however, had lower levels of essential fatty acids (low fat intake), vitamins of the B-complex group (cereals) and calcium (milk intake).

Another indication of the status and prestige attributed to various foods is indicated by those chosen as ritual food offerings to *deities* during *pūjās* (religious ceremonies). These ritual foods are usually *chang* or *jā̃ḍ*, rice, chickens or goats, eggs, *raksī*, ginger and butter. Furthermore, the various exchanges of meat play an important role in the maintenance of social relations. On many occasions, only men were present during the *pūjā*. The eggs, heart, gizzards and liver are first offered to the deity and later distributed, first to the shaman and then all other attendees. "Wild" green vegetables and other *jaṅgal* ingestibles are not offered to the deities because they are 'wild,' a part of their own terrain.

Historical evidence tells us that for centuries *jaṅgal* ingestibles have figured importantly in the lives of the monastic population. Milarepa (a Tibetan saint and poet, 1038-1122 AD) subsisted on "nettles" during his meditation and in Milarepa's biography, he said "although they are perfectly good food, have a stinging, hence, symbolically hostile, exterior" (Paul 1982:236). Joseph Rock (1930) reported that the monks of Radja *Gönpa* (3,250 m), Eastern Tibet, live in hermit quarters and subsisted on *tsampa* (T, *rtsam.pa* ,toasted ground barley flour), and boiled nettles in the summer. Subsisting on these nettles was considered the greatest sacrifice to their austere existence.

Alcohol

Alcohol (*chang, jãd* and *raksī*) is used for religious purposes and for hospitality by Sherpas and Rais. Alcohol can be a source of income, and is ceremonially used in arranging marriages, settling divorces, resolving disputes, and conducting transactions (Diemberger 1988). Events surrounding marriage for Sherpa women involves an extended process of rituals which end in the formalization of a marriage contract and the legitimization of children. At each of these rituals, large quantities of *chang* and *raksī* are consumed.

Guests are usually served *chang,* or *jãd* (in Rai households) and the cup is refilled as it is sipped. *Raksī* is served to special guests. With alcohol readily available, it is easy for a person to become drunk, which often happens. When Sherpas are travelling between villages, *chang* is carried and offered as hospitality by those travelling and also given by those in the house along the way. *Chang* is also given to children and I have seen Sherpa mothers use *raksī* to pacify a crying child.

Generosity is highly valued, both in ordinary social intercourse and as a source of religious merit. This sets the stage for ongoing cycles of reciprocity (March 1987). Ortner (1978:73) says, "it is clear that beer is used to manipulate others on the basis of its 'naturally corrupting powers'." The abundant supply of alcohol at a Sherpa wedding or a new house celebration ensures that monetary gifts will be generously given by those guests who attend.

Guru Rinpoche is credited with the invention of beer. "Beer serves as a ritual offering to God to gain protection" (March 1987:380,f22). The blessings of beer to Guru Rinpoche is very explicit. Barley flour is placed on the side of a china cup. Before drinking the *chang* (fermented grains) a person touches it three times with the fourth finger of their right hand and recites a few words as the blessing of beer to Guru Rinpoche. In March's (1987:381) translation of "The Tale of the Beer Yeast,"

Let's make good beer; do make the beer well. It's the blessing of the Guru; it's the blessing of power and strength of life.

Only women, and especially the head woman in each household, makes the yeast for brewing and brews the beer or distills *raksī* (March 1987). In the song, take the yeast made from all the right herbs, refers to the *titepāti* plant, -"let's make good beer- and moisten it with water" (March 1987:381).

Alcohol and "prestigious" foods are usually given to individuals of a high status as discussed earlier in this chapter. I observed however, among the Sherpas how two daughters-in-law used *jaṅgal* ingestibles to show respect to their father-in-law. One daughter-in-law brought *raksī* made of the *larewa* root (a *jaṅgal* tuber). Another daughter-in-law brought *sisnu* and rice one morning. Their father-in-law was very pleased and showed his approval of these gestures. This incident shows how behaviour differs from knowledge and attitude, and the intra-variability which occurs within the household varies with the particular situation.

The ways in which the religious and cultural elements of the foraging farmers influences how they use and manage the *jaṅgal* resources will be discussed further in Chapter 9.

CHAPTER 7

Women and *Jangal* Resources

In Nepal, culturally-determined gender division of labor within the household delegates responsibility to women for gathering and transporting fuelwood and fodder, carrying water, managing livestock and for essential aspects of food procurement, processing and preparation (Agarwal 1986; Dankelman and Davidson 1988 and Rodda 1991). An even more critical issue is that women are often the primary and, in female-headed households, the sole economic providers. Gender inequalities in access to productive and subsistence resources can make the role of women even more difficult (Agarwal 1991).

Deforestation changes the socioeconomic role of women as land degradation reduces the yield of cultivated crops and *jangal* resources and depletes firewood supplies. When time requirements to collect fuel increase, women find it ever more difficult to maintain subsistence levels in agriculture and still have sufficient time for food preparation and other domestic activities (Fernandes and Menon 1987; Kumar and Hotchkiss 1988). Due to the importance of fuel collection, Panter-Brick (1989) found that the state of pregnancy and lactation did not reduce the amount of the woman's participation in this task.

Although women have limited rights to private property resources such as agricultural land, communal resources have always provided women and their children of tribal, landless and marginal peasant households, "a source of subsistence unmediated by dependency relationships on adult males" (Agarwal 1991:20).

The location of available firewood, land tenure practices and seasonal variation, along with cooking styles, types of meals cooked, family size, and income-generating activities (dependent on firewood or other *jangal* resources) affect women's decision-making strategies. The enormous burden that women must bear has dire consequences. According to the 1981 Nepal Census, after one year of age, women had higher mortality rates and shorter life expectancies than men (UNICEF 1987). The 1990 Nepal census showed that women's life expectancy is still lower than that of men, 52.6 years for women, compared to 55.4 years for men (UNESCO 1991).

Adverse health consequences for the vulnerable groups arise from these factors:

- increased physical expenditure to collect fuel;
- changes in consumption patterns, such as a shift to less nutritious foods, a reduction in the number of meals cooked, partially-cooked food with a potential for toxicity, or a heavy dependence on low protein foods; and
- loss of dietary diversity available from the *jangal*.

Dankelman and Davidson (1988:71) discuss the close statistical association existing between per capita consumption of food with that of fuel; as fuelwood becomes scarce, nutritional status decreases. In addition to this, the importance of *jangal* resources to ensure nutritional adequacy is an important issue that warrants attention.

Women, pre-adolescent girls and children face more severe nutritional consequences from fuelwood shortages than men due to the often culturally determined intrahousehold food distribution practices (Agarwal 1986; UNICEF 1987). Adolescent and pre-adolescent girls who contribute heavily to family labor and are discriminated against in food distribution may be at greater risk (Acharya and Bennett 1981; Agarwal 1986; Dankelman and Davidson 1988; Leslie et al 1988; Pitt 1986). Last, I want to warn that the daughter-in-law, especially the newly arrived bride who has joined her affinal relations' household, and the woman married to the youngest son residing patrilocally, are often at greatest risk of social, health and nutritional deprivation.

Food Processing and Preservation of Grains

Before I discuss the processing of *jangal* ingestibles, it is important to describe the other food processing tasks which women perform on a daily basis. Women, overburdened with subsistence chores, are confronted with the difficult choice of how to allocate their time. I had never seen women work so hard to eke out a living. They endured difficult weather conditions in order to ensure food would be brought to the hearth to feed their families.

Harvesting, processing and preparation of food grains is very time consuming. Seed crops: barley, naked barley, millet, and wheat, are usually harvested by women with an iron sickle (*hasiyā*) where a handful of stems are cut at one time. Once harvested, the processing of the grain such as barley involves a complex process:

- First, the outer fibers of barley are burned, a procedure which makes the grain easier to thresh.
- The grain is then put in a long flat bamboo tray (*gumu*) and placed above the open fire to dry.

- Once dried, the grain is threshed near the household with a long heavy stick; and then sorted and winnowed.
- The grain is dropped into a hand grinder consisting of two stones (*jāto*) which is turned by hand until it is ground into flour.

Mineral complexes released from the stone querns may increase the calcium and iron content of the flour (Messer and Kuhnlein 1986). During the monsoon season when the river is high, the water mill is used to grind the grain, so labor normally expended for the hand grinder is saved. Another benefit of hand processing of grains and the use of the water mill is that it ensures the retention of thiamine. The deficiency of this vitamin often occurs in areas where rice is highly milled (Gopalan et al 1984).

Rai and Sherpa women spend an average of four to six hours a day around the smoky hearth preparing meals and beverages. The grains are made into *dhīro* (a paste-like mixture comprised of flour mixed with boiling water); *chang* and *jāḍ* or *raksī*. Maize is coarsely ground and boiled as the main meal. Wheat and naked barley flour is used to make unleavened flat bread. Potatoes are usually boiled, fried or made into *dhīro*. *Yö* (T, *Yös*) (popped naked barley and maize whole grain kernels) is a popular snack among all social groups.

Limbu girls using a *jāto* to grind maize into coarse flour.

Food Processing and Preparation of *Jangal* Ingestibles

Although there are no distinct gender differences in the collection of *jangal* ingestibles, once they reach the household, women assume the major responsibility for their flow in the food path. The extensive ethnobotanical knowledge possessed by Rai and Sherpa women is apparent in the elaborate processing techniques used to remove the toxic compounds found in the *jangal* ingestibles. The processing of *jangal* ingestibles require a lot of firewood during the monsoon season. However, this does not decrease their importance as a food source, especially when other food supplies are low. Rai women are more dependent on *jangal* ingestibles than Sherpas, which will be discussed in a later section of this chapter. They spend an average of four hours daily collecting firewood (twice as long as Sherpa women). The women mainly collect dead branches and twigs off the ground. Firewood collection conflicts with other labor demands because potato crops need to be harvested, maize planted and livestock tended, in addition to childcare and domestic chores.

Both Rai and Sherpa women use traditional methods to save fuel and reduce smoke. These methods are coincidentally the main features of improved cooking stoves. Firewood is placed on rafters over the open fire to dry, especially during the monsoon months when the moisture content of firewood is higher. The National Academy of Science (1980) estimates that by drying wood to a moisture content of 20-25 percent, the amount of wood needed for a given heating requirement is reduced by twenty percent or more.

One labor saving technique used by women is to place buckets outside their houses to collect rain water during the monsoon season. This saves time because as mentioned in Chapter 3, Rai women spend between one-half to three hours a day roundtrip to fetch water. I estimated the amount of time it took the women to walk from their households to the nearest water source. Some houses are close to the water source and others are quite far. Women frequently complained of cuts and aching joints suffered during falls on the slippery and treacherous footpaths.

Foods reserved for "times of scarcity" involve not only the energy and time of labor collection, but also require complicated and lengthy processing and preparation techniques. Cultural encoding of rules are used to detoxify the 'wild' tubers so they will be an edible source of nutrition. An example is the 'wild' root, *phi to* (*Arisaema flavum*) which is collected occasionally when time is not a limiting factor. *Phi to* contains a bitter toxin (*cyanogenic glycosides*), which is a common property of "famine foods." The toxin acts as a protection against predators such as herbivores, and is also said to be a selective trait favoring those people with biological resistance (Huss-Ashmore and Johnston 1994). I was the only person in the Sherpa household who experienced a burning sensation in my mouth after eating the *phi to* during the evening meal, which indicated I was sensitive to even a slight residue of the toxin.

Phi to is dug up by a Rai shepherd at the *lekh*.

Cultural practices are used to detoxify the *phi to* and convert inulin, a complex carbohydrate, to fructose and glucose, making it digestible (Kuhnlein and Turner 1991). These preparation techniques involve:

- The *phi to* root is washed and then cooked in boiling water an entire night. Open boiling hydrolyses *cyanogenic glycosides* into hydrogen cyanide and sugar. The hydrogen cyanide evaporates with the vapor (Standal, personal communication, 1997).
- The cooked *phi to* is then peeled, beaten on top of a rock, and rolled into long strips.
- The detoxified *phi to* is made into bread, added to soup, or used to make *raksī*.

Sisnu leaves (*Girardinia diversifolia*, *Girardiana palmata* and *Urtica dioica* L.) require a long cooking period to make the tough leaves edible. *Sisnu*, together with potatoes or corn and spices, is served as a main course stew by Sherpas. *Sisnu* leaves and other 'wild' greens are boiled and made into a soup which is cooked and eaten mainly by Rai women, but also poorer Sherpa women as well. This method prevents the loss of nutrients since the liquid is drunk. Salt and chillies, the only condiments used by the Rais, are added to their meals.

Kudo (S) is a candy made from 'wild' honey and barley. The honey is boiled until it becomes very thick and is then mixed with the grain. A Sherpa woman living in Dobatak was given the honey by a Yangden Rai 'honey hunter.' He had extracted the honey from bee hives located on the high cliffs above Saisima. The grubs in the honey comb provide "hidden" sources of iron and protein (Standal personal communication, 1997).

Food Preservation Techniques

Food preservation techniques are used to store foods for the *anikāl* period. Rai women make a fermented dried vegetable referred to as *"gundruk,"* out of kale leaves (*rāyo sāg*), radish leaves (*mulā sāg*), rape leaves (*tori sāg*) and *mangan sāg* (*Cardamine hirsuta*). *Gundruk* is made by first cleaning the leaves and leaving them in the sun for an entire day to wilt. The wilted leaves are then beaten in a mortar and pressed in a large container. After two to three days, the leaves are covered with boiling water and left in a sunny place. The fermentation process takes one to four weeks depending on the type of vegetable used. The vegetables are removed from the container and dried in the sun for about one week. *Gundruk* is given to Sherpas by the Rais as a gift because Sherpas do not make it.

Rais also make *mulā sinki* which uses the radish root. The radishes are harvested before they flower. The root is dried in the sun for one day. Then it is put in a bamboo container and pressed hard with an implement. The root is covered with leaves and soil is placed on the top. Afterwards, the container is set outside to sun dry. Lactic acid forms and gives it a sour taste. *Mulā sinki* is stored for the winter months and taken to the *lekh* by the Rai shepherds.

Two Sherpa women who watch high-altitude animals at the *lekh* during the monsoon collect *mangan sāg* (*Cardamine hirsuta*). They use a simplified procedure to preserve the *mangan sāg* by drying it in the sun for two hours. As they explained to me, this procedure removes the bitter taste from the *mangan sāg* so it becomes tasty. The bitter taste may be a protective component present in the 'wild' leaves and sun drying it helps to release the toxin. The *mangan sāg* is cooked after all their supplementary food supplies are consumed.

Mangan sāg drying in the sun.

Yeast starter (*marcā*) is made by Sherpa women using millet flour and the *titepāti* plant (*Artemisia vulgaris*). Yeast is needed in the fermentation process for making beer. The yeast starter is made by the following procedure. Millet flour is added to hot water and made into a flour dough. Small balls of this mixture are shaped and put on a bamboo tray. Two balls from a previous batch, which had been specially set aside are crushed and sprinkled over the top of the freshly made flour balls. The *titepāti* leaves are then sprinkled over the top of the balls and the whole mixture is left indoors for a few days until it turns a white color which indicates the starter has grown. The leaves are removed and the fermented flour balls are smoked dry over the open fire. *Titepāti* leaves have a dual purpose: they help to retain the moisture in the yeast balls and *titepāti* is an important religious plant described in Chapter 6. Twenty balls are bartered by the Sherpas for four kilos of maize with the Rais.

Several other preservation techniques are used to store food for the winter months. Sherpa women make butter, *chura* (*T, phyu.ra*) (small pieces of a kind of dried cheese) and *somar* (S. a smelly cheese with a strong taste) prior to the winter months and before milk production is at its lowest level. The Sherpa practice of saving dried cheese for the winter months is an important adaptive strategy. Dried cheese is very high in protein because it is made from buttermilk and represents 23 percent of the weight of the yogurt (Goldstein and Beall 1990). Rai and Sherpa women dry meat over the fire. Bundles of three to four ears of corn are tied together and hung from rafters to dry. This procedure prevents rats and insects from eating the corn.

Food preparation, processing and cooking carries health risks for women. In Nepal, chronic bronchitis is higher in women than men. Pandey (1984) has reported that the prevalence of chronic bronchitis is associated with the time spent near the cooking fire. Rai and Sherpa women spend many hours daily cooking over three stones (*chulo*) or a metal ring with legs (*odhān*) and are exposed to high carbon monoxide concentrations from the fuels. During the winter months, stoves are lit for space heating purposes which further exposes women to domestic air pollution. Rai women have an additional risk because they are heavy tobacco smokers.

Another study by Pandey (1984) showed acute respiratory infection (ARI) in infants and children to be one of the major causes of death in Nepal, and the morbidity rate for infants (birth to one year) is the highest ever reported in the world. The author further warns that an additional risk factor is caused by passive smoking parents which can lead to the development of ARI. The time spent near the stove was found to be a statistically significant risk factor for acute respiratory infection (Reid et al 1986). Smith (1997) claims that those children exposed to indoor air pollutants may have two to five times a greater risk of serious ARI than children less exposed. Smith, however, cautions that further research is needed to confirm these effects. Rai and Sherpa women breastfeed their infants or hold them on their laps while cooking.

Improved cooking stoves have been introduced throughout villages in Nepal with varying results (Pandey 1991). An improved *chulo*, a gift of the Woodland's Mountain Institute (renamed The Mountain Institute), was brought from Kathmandu for a 41-year-old Sherpa women who had been suffering from chronic bronchitis for many years. She was not able to work and would take care of other women's infants and young children. The stove was broken when it was carried along the rugged three-day walk from Khandbari. The woman's family was asked to make windows in their enclosed bamboo house. The family said they used the improved *chulo* with chimney for only one to two weeks, because compared to the traditional three-stone fireplace, it requires smaller pieces of firewood and takes a longer time to boil water. The stove does not warm up the household nor does it accommodate large pots needed for making *chang*. Another reason the family stated they returned to using the traditional stove was that the fire must be constantly tended, preventing them from doing other domestic chores while the meal was cooking.

Consumption of *Jaṅgal* Ingestibles

Data gathered from the household surveys, food frequency checklists, 24-hour dietary recall method and participant observations have been used to report the consumption of *jaṅgal* ingestibles by Rai and Sherpa women and their families. Even though the sample size is small, a combination of nutritional methods has helped to portray a more complete dietary picture. The sample size was too small for statistical purposes but revealed useful information regarding the availability and frequency of the use of *jaṅgal* ingestibles in the Rai and Sherpa households.

Household Survey

A household survey was administered to 32 Rai and 20 Sherpa households asking the interviewees to name both cultivated vegetables and *jaṅgal* ingestibles gathered during the winter months and the monsoon season.

Table 22

Cultivated vegetables and *jaṅgal* ingestibles gathered by Rais and Sherpas during winter

Utilization Patterns	Local Names	Description
(R, S, c)	*Rāyo sāg*	*Brassica juncea* (Leaf mustard)
(R, S, c)	*Mulā sāg*	*Raphanus sativus* (Radish leaves)
(S, c)	*Pīḍalu sāg*	*Alocasia indicum* leaves
(R, c)	*Tori sāg*	*Brassica napus* L. (Indian rape)
(R, w)	*Sisnu*	a) *Urtica dioica* L.(Stinging nettle)
		b) *Girardinia diversifol* (Himalayan nettle)
		c) *Girardiana palmata* (Himalayan nettle)
(R, S, w)	*Bhurmang sāg*	A 'wild' leafy green
(R, w)	*Ban tarul*	*Dioscorea versicolor*
(R, w)	*Bhyākur*	*Dioscorea deltoidea* (Deltoid yam)
(R, w)	*Gitthe tarul*	*Dioscorea bulbifera* (Air tuber)
(R, w)	*Ghar tarul*	*Dioscorea alata* L. (White yam)

Key: R-Rai; S-Sherpa; c-cultivated, w-wild.

During *anikāl*, which coincides with the winter months, the Rais in particular, depend on the consumption of *jaṅgal* ingestibles to make up deficiencies in their daily diets. The household survey revealed that during the winter season, even though cultivated vegetables are available, Rais still exploited a wide range of *jaṅgal* ingestibles to supplement their main staple of coarsely ground boiled maize. According to Huss-Ashmore and Johnston (1994), the collection of 'wild' greens is a low-cost strategy with a minimal commitment of household resources. Women often gather 'wild' greens growing in the same area where they are collecting fodder or performing other farming or herding activities.

The 'wild' tubers and potato yam are a staple substitute and buffer when nutritive stress occurs. According to Mabberley (1993), tubers can grow up to 50 kg. These 'wild' yams supply carbohydrates (energy), for example, *ghar tarul* (greater yam) provides 147 kilo calories per 100 grams, one and one-half times the energy available in potatoes (97 kilo calories per 100 grams) (Adhikari and Krantz 1989). If at least a kilogram is eaten, they also contribute a significant amount of protein, thiamine and Vitamin C to the diet (Coursey 1967). *Bhurmang sāg* was the only 'wild' ingestible collected during the winter months by the Sherpas. Another factor to consider in the collection of *jaṅgal* ingestibles is proximity. These *jaṅgal* ingestibles are found at a lower elevation in the vicinity of the Rai village, at a time when the Sherpa villages are covered with snow.

Table 23

Cultivated vegetables and *jaṅgal* ingestibles gathered by Rais and Sherpas during monsoon

Utilization Patterns	Local Names	Description
(R, S, c)	*Iskus sāg*	*Momordica charantia* (chayote)
(R, c)	*Pharsī sāg*	*Cucurbita maxima* (pumpkin)
(R, S, w)	*Sisnu*	a) *Giardinia diversifolia* b) *Urtica dioica* L. c) *Girardiana palmata*
(R, S, w)	*Gutel*	*Trewia nudiflora* L. (a bamboo species)
(R,S,w)	*Kālo nigālo*	*Arundinaria falcata* (a bamboo species)
(S, w)	*Niguro*	*Dryopteris cochleata* (fiddleheads)
(S, w)	*Phi to*	*Arisaema flavum* (a tuber)
(S, R, w)	*Mangan sāg*	*Cardamine L. hirsuta* (bittercress)
(S, R, w)	*Cyāu*	various mushrooms

Key: (R-Rai, S-Sherpa, c-cultivated, w-wild)

Tusa prepared for the main meal

Niguro and *phi to* are exclusively collected by the Sherpas during the monsoon. These *jaṅgal* ingestibles are found at a higher elevation above the Sherpa village. Both Rais and Sherpas collect mushrooms, 'wild' greens and *tusa*, a term used to refer to any type of young bamboo shoots. Fewer cultivated vegetables are available during the monsoon, and Sherpas told me they like the variety in their diet which the 'wild' ingestibles provide.

Food Frequency Checklist

A food frequency checklist was administered to 15 Sherpa and 24 Rai women. The Sherpa women were asked the frequency of consumption of specific *jaṅgal* ingestibles throughout the year according to the categories shown in Table 24.

Table 24

Frequency of consumption of *jaṅgal* ingestibles by Sherpa women

Ingestibles	d	3x	2x	wkly	mth	yr	nr
Sisnu	2	2	4	3	4	-	-
Tusa	2	5	1	-	5	1	1
Niguro	1	1	2	4	6	-	1
Phi to	-	-	-	-	-	6	9
Bhurmang sāg	-	-	4	6	3	-	2
Mushrooms	1	1	1	-	6	4	2

Key: d=daily, 3x=3 times weekly, 2x=2 times week, wkly=weekly, mth=monthly, yr=yearly, nr=never.

More than one-half of the Sherpa women eat *sisnu* and *tusa* two or three times a week during the monsoon period. Two of the poorest households reported they fetch *sisnu* and *tusa* daily. Many factors must be weighed in determining whether *jaṅgal* ingestibles can be gathered for the household, for example, the distance involved to collect it and whether able-bodied laborers have time. The Sherpa women said they eat *phi to* only once during the monsoon period because it grows high at the *lekh* and involves lengthy processing to make it edible. One woman living in Saisima said she collects *sisnu* daily only if she is staying at the cow shelter. Saisima is surrounded by a sacred forest and *sisnu* grows in disturbed areas located far from the village. In one female-headed household, there was "no older person" available to cut the *tusa*.

When I was visiting Rai households during the interviews, baskets of 'wild' greens were often visible along the bamboo walls. The Rai women identified them as *bhurmang sāg*. Nearly half of the Rai women said they eat the 'wild' greens twice a day when they are available. The food frequency checklist shows that Rai women eat more 'wild' leafy vegetables than Sherpa women. The majority of Rai women eat *sisnu* daily during the summer months and *bhurmang sāg* daily during the winter months.

Table 25

Frequency of consumption of *jaṅgal* ingestibles by Rai women

Ingestibles	d	3x	2x	wkly	mth (1x)	year	nr
Sisnu	19	5	-	-	-	-	-
Tusa	-	-	-	-	8	16	-
Niguro	-	-	-	-	6	12	6
Phi to	-	-	-	-	-	-	24
Bhurmang sāg	22	2	-	-	-	-	-
Mushrooms	-	-	-	-	11	11	2

Key: d=daily, 3x=3 times weekly, 2x=2 times week, wkly=weekly,
mth=monthly, year=yearly, nr=never

Meal Patterns (24-Hour Dietary Recall)

A 24-hour dietary recall was administered to all women of child-bearing age who had a child less than five years of age. Fifteen Sherpa women and twenty-four Rai women were interviewed, the average age of the former was 30 years (range of 21-44) and the latter 32 (range of 18-49). The women were asked what foods they had eaten during the previous 24 hours. The quantity of food items was not measured because this method can be intrusive to the woman and other household members. I was interested in the types of meals eaten and the quality of the diet rather than focusing on food portions in quantitative (energy terms). Meal combinations (staple and complementary food items, that is, side dishes) can help give a rough approximation of whether nutrients are available for family members. I was also spot checking to see whether *jaṅgal* ingestibles were included in the daily meal patterns.

Sherpa Dietary Patterns

The Sherpas eat four meals and snacks throughout the day. After rising in the early morning hours before sunrise, the family will eat a light meal. A pot of Tibetan tea is brewed for the morning meal. Family members then collect firewood, grind grain into flour and perform other household chores. The large meal is eaten around 10 a.m. before the family members leave to work in the fields or to herd the animals. The laborers either take a light meal with them to the fields or it is brought by the household mother during the late afternoon. The evening meal is also large and is eaten about 7 or 8 p.m., usually after all the fieldwork is finished. Sherpa women continuously snack through out the day while preparing food, visiting with friends or when guests drop by. The Sherpa women's meal patterns based on the 24-hour recall data and personal observations is shown in Table 26.

Table 26

The most common meal patterns for Sherpa women

Early morning meal:

Khole (a thin porridge of flour mixed with salt and either water, *chang* or buttermilk)

Leftover previous evening meal (*dhīro*, rice or potatoes) fried in butter or oil

Pustakari (*Phul tarul*)

Yö (popped toasted maize or barley kernels)

Palgi (S) (a soup of dried maize kernels, vegetables, beans, salt and chili, eaten during the winter months)

Main meal (eaten in the morning and/or in the evening)

Monsoon months:

Potatoes or *Pīḍalu* (*Alocasia indicum*) boiled and served with chili, garlic, salt and occasionally 'wild' garlic leaves (*zimbu*)

Dhīro (a paste-like mixture similar to *khole* but much thicker)

      Served with one of the following:

      Somar soup "smelly cheese," it has a smell similar to camembert cheese.

      Chayote or potato soup

      Soybean soup (soybeans are given by relatives in Dobatak, or as a gift from the Rais)

      Chicken soup (when there are rituals)

      Shoshim soup (S, *zho.zhim*) (made from the milk left in the churn)

      Shakpa (S. Sherpa stew made with wheat flour balls or rice; beans; garlic; butter; chili and salt)

Winter months

Maize (coarsely ground and cooked)

Dhīro served with either:

      Kale (*rāyo sāg*)

      Radish leaves (*mulā sāg*)

      Buttermilk

      Somar soup

Foods taken to the field

Boiled potatoes or boiled *Pīḍalu*

Yö

Chayote root

Roasted ears of corn

Satu (flour mixed with Tibetan tea)

Buttermilk or fresh milk (if at the cow shelter)

Beverages

Tibetan tea, *chang*, *raksī*, buttermilk or milk (cow shelter)

During the busy agricultural season, Sherpa women prepare *dhīro* as their main staple because the pre-roasted ground flour saves cooking time. Sherpa women stated that the other advantages of *dhīro* preparation is that it requires little firewood, is satisfying and is a favorite food. Meals are flavored with salt, chili, garlic, *timur* (*Zanthoxylum armatum* DC and *silam* (*Perilla frutescens* L.) which grows on millet fields.

The Sherpas drink *chang* daily and *raksī* about three times a week. Alcoholic beverages are estimated to provide 406 calories of energy and 3.5 grams of protein per kilogram. (Piazza 1986). This figure would vary according to the consistency of the beverage, that is, the amount of grain used in relation to water. Even though I was not able to find a nutritional analysis for these beverages, chang may supply minerals and other micro-nutrients to the diet. An analysis by the Central Food Laboratory in Kathmandu found the percentage of alcohol in *chang* to be 2-3 percent (Kunwar 1989).

Tibetan butter tea is drunk daily in most Sherpa homes. Tibetan tea is made from brick tea which is traded or bartered from Tibet or purchased in Kathmandu. The hot tea is poured into a churn. Then butter, salt and occasionally milk are added. The tea is churned thoroughly and then poured into a pot and reheated over the fire. The Tibetan tea mixture is similar to a soup broth. The tea cup is "bottomless" because it is continuously refilled. Only when the tea is drunk quickly and the cup pushed aside, does the hostess stop refilling it. The methyl xanthines present in the tea has been hypothesized to be beneficial in human adaptation to hypoxia in high-altitude regions (Larrick 1991). Although the tannins in tea inhibit mineral absorption, for example, iron, if milk is added to tea, the milk protein casein binds with the tannin and lessens the binding of the minerals with the tannins (Shils and Young 1988). Sherpas who own milk-producing animals may be getting the additional benefit of increased iron absorption, if they add milk to their tea.

The majority of Sherpas do not rely on *jaṅgal* ingestibles as much as Rais because they own a greater number of livestock. The ownership of high-altitude animals is synonymous with wealth, which is apparent by the nutritional contribution of these animals to the family's dietary intake. Milk production varies among different animals. The Sherpa herders said that "the *caũrī* (yak-cow hybrid) produces twice as much milk as the *nak* (female yak)" This was also noted by Downs (1964). Stevens (1993) pointed out that the higher milk yields for *caũrī* may reflect differences in the greater abundance of fodder in areas where they are grazed compared to the higher altitude where *nak* are grazed. Table 27 shows the milk production and percentage of fat for high-altitude animals and cows (Joshi 1982).

Table 27

Milk production and percentage of fat for high-altitude animals and cows

Animal	% Fat in Milk	Average Milk Production (kg/day)
Nak	5.5- 9.5	4
Dimjo caũrī	4.5- 7.0	6.5
Urang caũrī	4.5- 6.5	5.0
Pamjo caũrī	4.5- 6.5	5.0
Pamu	4.5- 6.0	3.5
Tolmu	4.5- 6.0	2.5
Kirkho cow	4.5- 5.0	2.0
Hilly cow	3.5- 4.5	1.0
Zebu cow	3.5- 4.5	2.0

Key: *Nak*= female yak, *dimjo caũrī*, a crossbreed of a *kirko* bull and a *nak*; *urang caũrī*, a crossbreed of a yak and a cow, *pamjo caũrī* =a *kirko* bull and a cow; *pamu*= a Tibetan *kirko* cow, *tolmu*= a *caũrī* and a *kirko* bull, *kirkho* cow= Bos taurus, *Zebu* cow= Bos indicus. Key is based on Brower (1991) and Stevens (1993).

The butterfat content of *nak/caũrī* milk was seven to nine percent in the analysis by Schulthess (1967). The percentage of fat in milk can vary seasonally. It is greatest during the months of September-November and is reduced between the months of June and July. Both Joshi (1982) and Schulthess (1967) did not state the season of the year when the fat in the milk of the various animals was analyzed.

High-altitude animals produce milk and milk products (cheese, yogurt, buttermilk, and dried cheese), which are important sources of protein. Salt is added as a preservative to *nak* milk, yogurt, butter and cheese. The amount of a household's butter supply depends on the number of milk-producing *nak* owned. One *nak* produces milk to make about 0.5-1 kg of butter per month. Butter is harvested throughout the year with a peak in mid-summer. Sherpas live at a higher altitude than Rais, about 2,300 m, and require additional calories to maintain an adequate production of body heat to prevent hypothermia, especially during the cold winter (Baker 1973). The addition of butter to the Sherpa's salted tea adds energy to their diet. Butter has nine calories per gram (more than twice as much as carbohydrates).

A Sherpa woman milking a *caŭrī*.

The diet of vulnerable groups (children and women of child-bearing age) may only be adequate when milk-producing animals are at the base village and prior to their departure for high-altitude pastures. Casimir (1988:354) asserts that if buttermilk is included as a "regular" beverage (implying large quantities are drunk daily) in a person's diet for about six months a year, calcium can be stored in the skeleton; Vitamin C in the viscera, particularly the liver; and Vitamin B2 (riboflavin) "ubiquitously" in all body tissues during the winter months when dairy products are reduced.

The Rai Meal Pattern

Upon rising in the early morning hours, the Rais leave to fetch firewood or water without partaking of any food or drink. The Rais dietary pattern usually consists of a large meal in the late morning between 10 and 11 a.m. The staple food most frequently consumed is *chakla* (boiled coarsely-ground maize). It is a superfood (McElroy and Townsend 1985:191) for the Rais because they are culturally and economically focused on it as their main cereal staple. Side dishes, mainly vegetables, are usually cooked with salt and water, and chili is added, if available. Chili provides minimal amounts of Vitamin A and ascorbic acid. Garlic, given as a gift by the Sherpas is used to flavor their meals as long as the supply lasts. In the evening, around 7 p.m., an identical meal may be served. Table 28 shows the following meal patterns for Rai women based on the 24-hour dietary recall and personal observations.

Table 28
The most common meal patterns for Rai women

Main morning or evening meal:
Millet or maize *dhīro*
Chakla - most common
      Served with:
            Sisnu (Himalayan or stinging nettles)
            'Wild' leafy green (*bhurmang sāg*)
            Soybean soup
Chicken soup (yearly livestock *pūjā* ritual)
      A soup made of radish leaves
      A soup made of kale leaves
      Gundruk soup

Poorer families may have only one of the following as their meal:
Bhurmang sāg soup only
Sisnu soup only
Maize *dhīro* with salt and chili only
Maize or millet *khole* only

Bhurmang sāg (a 'wild' green) was eaten by most Rai villagers with *chakla* twice a day during the winter months. Not all the households reported consuming a staple food at their meals. Hence for the Rai farmer who has small landholdings and few livestock, household food supplies are insufficient for the household members. A few women appeared distressed when they said that their meal had consisted of only 'wild' greens made into a soup. The normal practice is for 'wild' greens to be eaten as an accompaniment to the main staple because if they are eaten alone, they "will not fill the stomach."

The method of meal preparation used by the woman can positively or negatively affect the nutritional status of her household members. I noticed both Rai and Sherpa women washed the vegetables prior to cutting them which helps to prevent a loss of nutrients. They used cooking methods which enhanced nutrient retention in the prepared meals. Leafy green vegetables are often made into a soup which minimizes the loss of nutrients. The pots are covered with a lid which lessens the food item's exposure to air and prevents the loss of ascorbic acid. The boiling of unpeeled root vegetables by both Rai and Sherpa women helps to prevent the leaching of nutrients into the water. Also, the use of cast iron pans for frying vegetables, soybeans or other foods, can increase the iron content of the diet.

I observed in one Rai household at the morning meal that whereas all family members ate the *chakla*, the complementary food for the adults was a soybean soup and the young children were given milk. Not all Rai families have milk and it is altruistic that the milk was channelled to the young children. Once soybeans are finished, Rai women told me they depend on 'wild' greens from the *jangal*.

During the dietary interview, Rai women said they only eat two meals a day, and did not mention food (snacks) eaten between these meals. In many cultures, snacks are not viewed as "food." The importance of snacks or between-meal foods cannot be taken lightly. Although the Rai women did not report any snacks on the 24-hour dietary recall, I know, however, that they often snack during the day between the main meals. Personal observations in Rai households revealed bananas, cucumbers, sugarcane, fried soybeans, roasted corn and 'wild' berries were common snacks. If snacks such as *gitthe* (*Dioscorea bulbifera*), a handful of roasted soybeans and 'wild' raspberries were consumed, their dietary choice would be satisfactory in providing nutrients needed for health and maintenance.

As mentioned in Chapter 3, I did not use the food weighing method because I did not want to intrude or alter the normal eating pattern of the household members. For this reason, the nutrient adequacy of the mother's diet can not be determined, nor can the amount of nutrients be measured. In contrast to Rai women, Sherpa women eat continuously throughout the day, as shown in Table 26. Sherpa women appear much larger than Rai women, and a fat appearance is equated with strength and the ability to carry heavy loads, a quality sought in a prospective bride.

The 24-hour dietary recall and the food frequency list revealed that Rais drink mainly water between meals. *Jãḍ* (the Rai's local fermented beverage similar to *chang*, but usually of a much thinner consistency) or *raksī* (distilled liquor) is mainly served for special celebrations. Rais generally have less grain and for this reason their *jãḍ* is of a thinner consistency than the Sherpa's *chang*, which may lessen its importance as a source of energy and B vitamins. The Rais are known for their *raksī*, which is considered strong by the Sherpas. The strength of *raksī* is determined by the number of times water is added to the grain, the less water, the stronger the alcohol. To make one liter of *raksī* about four kilograms of maize, millet or other grain is needed.

A greater variety and frequency of *jangal* ingestibles are eaten than revealed by the dietary survey methods. On my journey with the Rai shepherds to the *lekh*, I saw them picking a number of 'wild' leafy greens, *molang sāg* and *mangan sāg* (*Cardamine L. hirsuta*) on the high-altitude pastures. These 'wild' greens in addition to *gundruk* and *mulā sinki* (see Chapter 7) are carried to the lekh, as supplementary foods to the maize meal. On a few occasions in Sherpa households in early spring, I was served *jaringo sāg* (*Phytolacca acinosa* Roxb.).

Nutrient Composition
Protein

The types of meals were observed in order to determine whether the combination of the main staples (various grains and potatoes) with pulses and vegetables would ensure a complete protein source. In the absence of milk products, the right combination of grains and pulses can provide all eight essential amino acids. For example, 30 grams of legumes is sufficient to provide enough protein when the main staple is barley or potatoes. Soybeans, in addition to providing protein, is high in certain B vitamins, calcium, phosphorus and iron (Simoons 1991). Another benefit of combining cereal grains with pulses is that the former provides the amino acid methionine and the latter, lysine, which the other is deficient in, and the combination of these foods will make a complete protein (Adhikari and Krantz 1989).

Maize is deficient in the amino acids lysine and tryptophan which are needed to make niacin. By combining maize with other niacin-rich foods at the same meal, for example, millet, barley, oil seeds, pulses, or *gundruk,* protein requirements are met. Rais often eat both *chakla* and millet *dhīro* at the same meal ensuring an adequate amino acid balance. Millet contains very good quality protein, is rich in B vitamins and among all the cereals has the highest calcium content (Adhikari and Krantz 1989). Leafy green vegetables generally contain 2-7 percent protein (fresh leaf values) which can significantly contribute to protein intake when eaten on a regular basis (Adikari and Krantz 1989). The Central Food Research Laboratory, His Majesty's Government of Nepal, Ministry of Agriculture, is mainly concerned with the analysis of foods in Nepal. Only some of the known *jaṅgal* ingestibles have been analyzed because the procedures are very expensive and time-consuming. Another problem is that the local names for the *jaṅgal* ingestibles vary throughout Nepal by ethnic group, and geographical area. Thus several of the *jaṅgal* ingestibles, named in the Rai and Sherpa languages, may have already been identified and analyzed. The nutritional value of *jaṅgal* ingestibles in the following table have been taken from Adhikari and Krantz (1989) and Gopalan et al. (1984).

Table 29

Amount of protein in cultivated and 'wild' vegetables

Vegetables	Protein (gms/100 gms)	Moisture (gms)
Sisnu (stinging nettles) (w)	6.9	
Colocasia leaves, black (*)	6.8	78.8
Rape leaves (c)	5.1	84.9
Pumpkin vines (c)	4.6	81.9
Sweet potato leaves (c)	4.2	80.7
Mustard leaves (*rāyo*) (c)	4.0	89.8
Colocasia leaves, green (c)	3.9	82.7
Bamboo shoots (w)	3.9	88.8
Radish leaves (c)	3.8	90.8
Niguro (ferns) (w)	3.4	
Garlic leaves (c,w)	2.4	

Key: c= cultivated, w= 'wild', * not known whether c or w.
Source: Adhikari and Krantz 1989; Gopalan et al 1984.

Sisnu was the richest source of protein of all the vegetables, both cultivated and 'wild'. *Ban silam* (*Perilla frutescens*) is an herb which grows in swidden fields and its seed is used as a side dish condiment. A 100 g portion supplies 15.7 g of protein. However, even though the intake of a person would be a lot smaller, it is still an important source. 'Wild' and domesticated roots and tubers have also been analyzed for protein and are shown in Table 30.

Table 30

Amount of protein in 'wild' and domesticated roots and tubers

Roots and tubers	Protein (100 grams)	Moisture (grams)
Ghar tarul (w)	4.1	79.6
Gitthe (w)	3.7	
Colocasia root (w)	3.0	73.1
Yam, 'wild' (w)	2.5	70.4
Small onion (c)	1.8	84.3
Potato (c)	1.6	74.7
Yam, ordinary (c)	1.4	69.9
Sweet potato (c)	1.2	68.5
Carrot (c)	0.9	86.0

Source: Adhikari and Krantz 1989.

'Wild' roots and tubers can be an important source of protein (depending on the quantity eaten). All the varieties of 'wild' roots and tubers outweighed cultivated tubers in the amount of available protein.

The *jaṅgal* provides important riverine and insect sources of protein often missed in dietary surveys. Fish, also provides omega-3 fatty acids and fish bones, if eaten, are a rich source of calcium. Rais fish in the Apsuwa *Kholā* and make a fish soup as an accompaniment to the maize meal. One young Sherpa boy pointed to a hornet nest high in a tree and said, "At night the nest is burned, and later the hornets are fried and eaten."

The contribution of foods high in protein must be understood in relation to the daily dietary requirements which are being used for the Nepalese people. According to the 1981 Recommended Dietary Intake for Indians, 45 g of protein are required for women, 59 g of protein required for pregnant women and 70 g of protein required for lactating women (Gopalan et al 1984). The requirements are often not met by poor farmers, who may be more prone to infections and less resistant to diseases.

Vitamin C

Potatoes, radishes, 'wild' and cultivated green leafy vegetables and 'wild' fruits, are an important source of ascorbic acid which enhances the absorption of the non-heme iron present in a principally grain diet. The 'wild' fruit referred to as *amalā* (*Phyllanthus emblica*), located in the lower region and commonly eaten by the Rais, is one of the richest sources of ascorbic acid (Gopalan et al 1984). The 'wild' fruits are usually "eaten on the spot," hence the ascorbic acid present in the fruit is not destroyed by processing techniques (heating or drying). The amount of ascorbic acid, however, decreases during the long-term storage of these foods (Gopalan et al 1984). 'Wild' chives (*zimbu*) are a high source of Vitamin C. Kuhnlein and Turner (1991) assessed the Vitamin C in 'wild' chives (*Allium Schoenoprasum*) at 32 mg/100g fresh weight.

An important consideration is that if 'wild' chives and fruits are consumed in small quantities, they may not contribute a large enough portion to meet the daily recommended requirement for ascorbic acid and other ascorbic rich foods would need to be included in the daily diet. Another important consideration is the seasonal occurrence of these *jaṅgal* ingestibles. The following table shows the amount of Vitamin C in vegetables and tubers.

The recommended amount of ascorbic acid set for pregnant and non-pregnant women by the Indian Council of Medical Research (1981) is 40 mg (Gopalan et al 1984). However, lactating women require double this amount. Turnips are an important source of Vitamin C, more than twice the amount found in potatoes. All of the 'wild' tubers have not yet been analyzed for Vitamin C by the Central Food Laboratory. However, because they are potato-like and potatoes are high in ascorbic acid, it is possible that they may also be an important source of Vitamin C during the winter months when few other food sources are available.

Table 31

Amount of Vitamin C found in vegetables and tubers

Food	Vitamin C (40 gm/100g)
Vegetables	
Turnip greens (c)	180
Radish greens (*mulā sāg*) (c)	81
Rape leaves (*tori sāg*) (c)	65
Sisnu (stinging nettles) (w)	53
Garlic leaves*	36
Mustard leaves (c)	33
Sweet potato leaves (c)	27
Colocasia leaves, green*	12
Pumpkin vines (c)	11
Roots and tubers	
Turnips (c)	43
Sweet potatoes (c)	24
Potato (c)	17
Radish (pink) (c)	17
Radish (white) (c)	15
Potato yam (*gitthe*) (w)	6

*Not known whether 'wild' or cultivated
Source: Adhikari and Krantz 1989.

Casimir (1978) conducted research among Afghan pastoralists and found buttermilk to be an important source of ascorbic acid. Casimir's findings showed that Afghan adults drink about 3 kg of goat and sheep buttermilk daily. On the other hand, because Sherpa herders may only drink one to two dishes of buttermilk (600 milliliters) several times a week, other sources of ascorbic acid are needed to help them meet their daily requirements for it.

Vitamin A

Vitamin A is necessary for the formation of epithelial tissues and normal vision. I saw no cases of night blindness[36] nor Bitot Spots among any of the children in neither the Rai, nor the Sherpa villages. My observation of a lack of Vitamin A deficiency in the research area tends to agree with earlier findings reported by UNICEF (1987) and Worth and Shah (1969) for the mid-hill and mountain regions. As noted in Chapter 1, Jumla, a mountainous area in Western Nepal, was an exception because a high prevalence of xeropthalmia was found among children (National Vitamin A Workshop 1992).

In Baidya's (1991) analysis of cultivated and 'wild' sources of beta-carotene, 'wild' greens and traditional semi-domesticated vegetables were shown to have the highest concentration of beta-carotene. The vegetables were analyzed by the high pressure liquid chromatography method for carotene analysis at the Central Food Laboratory in Kathmandu. The values of beta-carotene for these vegetables are shown in Table 32.

Table 32
Amount of beta-carotene in 'wild' and cultivated foods

Crops/Food Plants	Beta-Carotene Level (ug/100g)
Stinging nettle (*Sisnu*) (w)	4595
Dried red chili (c)	4228
Pudina mint (*Mentha aquatica*) (w)	3677
Turnip leaves (*B. rapa*) (c)	2316
Carrots (c)	1890
Rape leaves (*rāyo sāg*) (c)	1785
Mustard leaves (*tori sāg*) (c)	1604
Radish leaves (c)	1301
Fern shoot (*Dryopteris sp.*) (w)	565
Pumpkin leaves (c)	497

Source: Baidya (1991).

Radish leaf rape *gundruk* contains 640 mg of pro Vitamin A and rape leaf (*tori*) *gundruk* contains 1,520 mg of pro Vitamin A. Gopalan et al (1984) asserted that approximately 50 g of leafy green vegetables in the diet will furnish an adequate

---

36 Night blindness occurs when the child does not see well at dusk. Bitot spots resemble fish scales which appear on the conjunctiva of the eye. Both night blindness and bitot spots are signs of Vitamin A deficiency.

amount of Vitamin A for adults as well as children. If an excess of Vitamin A is consumed when it is plentiful, what is not needed can be stored in the liver for the lean season when sources are not available.

Gopalan et al (1984:13) pointed out that cow's milk has seasonal variability which is directly related to the "succulent green grass rich in carotene" that the animals forage on. During the monsoon period, the Sherpa women often described the *lekh* grass as "making the animal's milk come." Cow's milk is richer in Vitamin A than buffalo milk, as evidenced by the yellow color of the butter, and *caũrī* (yak-cow hybrid) butter may be even higher. The liver oils of some species of fish are the richest natural source of Vitamin A (Gopalan et al 1984). The Rais frequently fish in the Apsuwa *Kholā* and the *asala* (Snow trout) may be another "hidden" source of Vitamin A.

In a longitudinal, interdisciplinary study assessing food consumption and nutritional status in a mountainous area of Nuwakot District, Central Nepal, a food weighing method was employed throughout the year and revealed that 41 percent of the total quantity of leafy vegetables consumed by the population were 'wild' and 59 percent were cultivated. Between February and June no cultivated vegetables were available and the 'wild' vegetable (*Galinsoga cilata*) was said to appear "as a welcome weed in the maize fields around April" (Koppert 1986). Koppert said that although vegetables are available in sufficient quantities, cultivated vegetables "are only available in reasonable quantities from the end of the monsoon period till January." Koppert (1986: 221-222) then remarked, "In terms of development policy one should remember the important place of the need for tasty vegetables throughout the year." The population has sufficient "tasty" vegetables to provide a year round Vitamin A source, however, the author, a nutritionist, seems to be disregarding the importance of the 'wild' greens to the local diet, which are consumed when cultivated vegetables are not available. This is an example of the cultural views held by some scientists and researchers which can enforce policies which may cause a deterioration in the nutritional content of the local diet.

Yellow maize, the main staple of Rais, is the only cereal that contains carotenoids, a source of Vitamin A (Widdowson 1991). A very low fat intake or high parasite infestation can impair Vitamin A absorption (McGuire 1993). The sources of fat for the Rais will be discussed in the section on fat.

Riboflavin

Caldwell and Enoch (1972) found 'wild' leafy vegetables had high contents of riboflavin (0.4- 1.2 mg/100 g edible portion), greater than that available in whole eggs, milk, nuts and fish. This is of utmost importance for Rais because 'wild' leafy vegetables may be their only source of riboflavin if they own few livestock. 'Wild' yams contain 0.47 mg/100 g of riboflavin (Adhikari and Krantz 1989). The Indian requirements used for Nepal has been set at 1.5 mg per day for pregnant women and 1.6 mg for lactating women.

Buttermilk is also an important source of riboflavin (Casimir 1978). The consumption of buttermilk thus enhances the nutritional content of the Sherpa diet.

Iron

All leaves, fresh and dried, have a high iron content. A 100 g portion of radish leaf (*rape*) *gundruk* contains 95 mg of iron. Other good sources of iron are berries, soybeans and millet (Adhikari and Krantz 1989). Gopalan et al (1984) assert that approximately 50 g of leafy green vegetables can help fulfill a major portion of the iron requirement for adults and children. Several times, I measured the raw weight of *jaṅgal* ingestibles before they were cooked. On one occasion, 440 g of *sisnu* was prepared as a side-dish for seven household members. Therefore at least a 60 g portion of *sisnu* would be available to each person. The iron-rich food sources are especially important for pregnant and lactating women who suffer from anemia. The consequences of this will be discussed further on in this chapter.

Green leafy vegetables are rich in Vitamin K which is necessary for proper blood clotting and prevention of bleeding. They are also good sources of folic acid, a deficiency of which can result in anemia and a widespread nutritional problem occurring among infants and pregnant women (Gopalan et al 1984) and Spina bifida problems in newborns (Standal, personal communication, 1997). Fortunately, during pregnancy green vegetables, considered to be "cold foods," are not restricted; thus the mothers are not deprived of their beneficial health and nutritive properties. The following table shows the amount of iron in 'wild' and cultivated vegetables.

Table 33

Iron content in 'wild' and cultivated tubers

Tubers	Iron (mg/100g)
Ghar tarul (w)	24.0
Rani bhyākur (w)	8.4
Carrots (c)	2.2
Yam, ordinary (c)	1.3

Source: Adhikari and Krantz 1989.

The 'wild' root (*ghar tarul*) has nearly three times the amount of iron available in the other vegetables. A 100 g portion of *ban silam* contains 11.1 mg iron; however, the amount consumed by a person would be much smaller.

Calcium

One of the important functions of calcium is its role in blood clotting, in addition to its more commonly recognized role in the formation and building of strong bones and teeth. Children and pregnant and lactating women have an increased requirement of calcium.

Table 34
Calcium content of 'wild' and cultivated
vegetables, roots and tubers

Vegetables	Calcium (mg/100g)
Sisnu (stinging nettles) (w)	928
Colocasia leaves, black *	460
Rape leaves (c)	370
Sweet potato leaves (c)	360
Radish leaves (c)	265
Colocasia leaves, green *	227
Mustard leaves (c)	155
Roots and tubers	
Carrots (c)	80
Ghar tarul (white yam) (w)	70
Radish, pink (c)	50
Onion, big (c)	47
Onion, small (c)	40
Colocasia root *	40
Yam, ordinary (c)	35
Yam, 'wild' (w)	20
Gitthe tarul (air potato) (w)	12

Key: c= cultivated, w= 'wild', *= not known if c or w.
Source: Adhikari and Krantz 1989.

Sisnu (stinging nettle) is the highest source of calcium, especially important among families who own few, if any, milk-producing animals. According to Adhikari and Krantz (1989:145), the drying and fermentation of leaves is a beneficial nutritional practice. The processing does not affect the mineral content, although Vitamin A and C are somewhat reduced. For example, 100 g of radish leaf (rape) *gundruk* contains about 2,500 mg of calcium, The amount of *gundruk* made into a soup is small, however, the contribution can still be significant. A 100 g portion of *ban silam* which is dried and used as a flavoring for food, contains 350 mg of calcium.

If taro corms are consumed in large quantities, they supply calcium, phosphorus, iron, thiamine and niacin (Murai et al 1958). A major hindrance to the absorption of calcium from the taro corms is the presence of oxalic acid which binds the calcium to form insoluble calcium oxalate, thus making it unavailable for absorption (Gopalan et al 1984). According to Shils and Young (1989), pokeweed (*joriṅgo sāg*) contains large amounts of oxalic acid. The presence of phytin, a substance found in cereal-based diets, also interferes with the absorption of calcium. These points need to be considered when one is assessing the nutritional quality of the diet.

Fat

Rai women make an oil out of the seeds of *philiṅgo* (*Guizotia abyssinica*, Niger seed), the seeds of *tori* (*Brassica campestris* var. toria Duth & Full.) and occasionally eat soybeans (20 percent fat) when seasonally available. *Shinik* (N), *lenza* (S), another seed plant made into oil, is found near Yangden. Nuts have a high fat content in the range of 40-60 percent (Adhikari and Krantz 1989), and walnuts (*Juglans regia* L.) referred to as *okhar* are consumed largely by Rais. Another source of fat which the Rais consume is *Patle* (*Castanopsis hystryx.*). *Patle* has a higher carbohydrate and lower fat content compared to walnuts and is similar to chestnuts (Standal, personal communication, 1997).

Fat appears to be a limiting factor in the Rai diet since the women said vegetables often boiled with only salt are frequently served as the accompaniment to the maize meal. Fat, besides being necessary for Vitamin A absorption, is an energy source. If the amount of fat in the diet is low, the protein available in the diet may be used as energy and neither for growth, nor the building and repair of tissues. The Rais place a high value on pig fat. At Rai weddings, fatty pieces of pork are put on leaves and distributed to all the guests as a special gift when they give monetary contributions to the wedding couple. The "special gift" may be an important adaptive trait to provide extra calories in the diet because fat sources in the Rai daily diet are minimal. The Hindu festival, *Dasaĩ*, which is celebrated during the month of October, and funerals are other occurrences when pork may be served.

Social Factors Affecting the Health and Nutritional Security of Women

Even though women are responsible for their families' welfare, their own nutritional security and well-being is often neglected by them and their societies. These adverse consequences can be seen in their life expectancy at birth, which as noted earlier in this chapter, is lower than males. Nutritional security differs from household food security because while household food security is concerned with the availability of food to the household, nutritional security is the result of the quality and quantity of food items available and distributed to each of the members (Zeitlin and Brown 1992). This illustrates how culture can influence opportunities and the potential for the success or failure of dietary strategies within the household. Meeting nutritional and energy requirements is very difficult, especially for pregnant or lactating women and children (Agarwal 1985). The bias in intra-household allocation of food, increased women's workloads and cultural views towards food are social factors which affect the nutritional security of these women (Zeitlin and Brown 1992).

Intra-Household Food Distribution

Gender inequalities in intra-household food distribution (IHFD) vary cross-culturally (Basu et al 1986). The study by Basu et al (1986) found no male bias in IHFD among tribal groups in India although it existed among the high economic Hindu caste of southern West Bengal. The authors concluded that gender inequalities in IHFD may be influenced more by ethnic and sociocultural, than economic factors. However, the authors warned that the results are based on a small sample size and should be viewed with caution. Another problem may be that the recall method was based on a one-day semi-quantitative data. An evaluation of only one day may be atypical not accounting for variation which occurs seasonally and during major feasts and celebrations.

As people become "Sanskritized" (Golpaldas et al 1983a:217), their diet becomes less diversified. The study by Golpaldas et al (1983a) of two culturally-different segments of the forest-dwelling group in Gujarat, India, showed that Sanskritization led to a rejection of many edible forest foods for children which resulted in them having inadequate intakes of Vitamin A, iron and ascorbic acid, a lower hematological status, and a greater prevalence of Gomez Grade III malnutrition.

Food is a symbol of social and religious value that influences how it is allocated within the household among family members (McGuire and Popkin 1990; Wheeler and Abdullah 1988). Gender bias in food distribution is related to factors such as age, birth order and sex of the child relative to siblings, maternal education, economic status and the economic contribution of household members (Gitteltsohn 1991; Zeitlin and Brown 1992). The allocation of food, the amount, type and order of serving, provide insights regarding the status of individual household members.

Patterns of power and authority within Sherpa society are visible at meal times around the family hearth. Meat is regarded as a highly prestigious food item for the Sherpas. Gender differential nutrition is linked with the high status of males who are given priority in meat portions and larger meals. The choicest pieces of meat are distributed first to the senior household male, usually a large dish containing the muscle and organ meats (high in protein, iron and vitamin B12) for him to taste and show his approval. Other household members receive a smaller serving of the meat item with their meal according to the following hierarchical order, other adult males by seniority, followed by both male and female children and older adult females according to age, and whether they are kin or affinally related.

The "Powerless" Daughter-in-Law

The fate of the girl child in Nepal had been brought to attention during the Centre for Development and Population Activities (CEDPA) Regional Conference on Women's Leaders held in Kathmandu in 1987. A resolution was proposed to declare 1988 as the 'Year of the Girl Child and Young Women' (Singh 1989). The amount of discrimination a girl child faces varies by ethnic group and according to the stage in her life cycle. A woman's position within the household and the quality of household interactions changes throughout her life cycle which has a bearing on her own health and nutritional status. A daughter has no right to parental property, except for when she has not married by the age of 35 years and remains unmarried thereafter (Singh 1989). Also, if there is no son in the family, her husband will reside matrilocally.

In my research area, I observed that young women, that is, the daughters-in-law – especially the youngest and most recent arrival to the patrilocal residence – are in a very vulnerable position and encounter social, health and nutritional deprivation. Overall, the wife of the youngest son has the lowest status in the household. She has the most work assigned to her by the household mother, and only upon the death of either her mother-in-law or father-in-law can she assume the powerful position of the household mother.

Her husband (youngest son) cannot split from his parents, but must care for them when they are old and carry out their death ritual. He usually only receives his property upon the death of one or both parents. This tradition can deviate from the norm and is changing with modernity. In the Sherpa village of Walung, a two-day walk northeast of Gongtala, the youngest son and his wife split from his parent's household and set up their own household nearby.

The daughter-in-law faces a dilemma, that is, whether to voice her concerns about her subordinate position, heavy workload and long hours, or remain quiet so as not to displease her mother-in-law and face even harsher work conditions and a meager food allocation at mealtimes. On a visit to a Sherpa household in another village, I remember observing how thin the youngest daughter-in-law was even though she was five months pregnant and had a toddler on her lap while cooking the household meal. The daughter-in-law faces the pressure of reproduction by her affinal relations. One young bride went trekking and kept her own money, angering her mother-in-law and not following the tradition that had been set out for her of staying patrilocally and bearing children.

Diemberger (1992:CH7, p.16) remarks that Khumbo women have to follow the same path: "from strangers to 'landladies,' from object dealt with in the alliances to producer of alliance strategies, ... for motherhood, symbolized by her blood, is thus at the same time her weakness and her power." Women cannot transmit the 'bones' symbolizing the patrilineal line and are mobile, which legitimizes "the general rule of patrilocality and the subordinate position of the daughter-in-law" (Diemberger 1992:CH7, p.24).

The youngest daughter-in-law eats last in the hierarchical order and often must cook additional foods for herself. If meat was served and the pot is empty, a non-meat food item is substituted. Gitteltsohn's (1991) research on intrahousehold food behavior in mid-Western Nepal found "adult women were less likely to meet their nutrient requirements for energy, beta-carotene, riboflavin, and Vitamin C," compared to adult men. Women's late position in the household serving order, the channelling of special foods to men and children and women's lower total intake of food accounted for this difference.

I noticed a more equitable distribution of meat in Rai households. One day during the winter months in Yangden, my research assistant and his younger brother had cut up a cow that had died because of the cold weather. Once the beef had cooked, my research assistant's father (the household head) dished up ladles of the beef soup into dishes. The meat portions were very generous and looked similar in size. He later gave seconds to everyone, regardless of gender, until all the soup had been served. The men and I each got a dish for the *chakla* (maize staple) and the *dhīro* (a boiled flour mixture). The women's food was served on leaves because there were not enough dishes. Later *jā̃ḍ* was served. I observed that when Rais serve meat, a larger portion is given, while Sherpas make more of a soup out of it to stretch it further.

An unintentional inequitable allocation of protein food does occur when men are away from the home village. I observed this while travelling with the Rai shepherds when they were moving their sheep to the *lekh* for the summer months. It is a common occurrence for a sheep or other animal to die from illness or injury during the 7-10 day journey. The men have the prerogative of choosing which parts of the animal will be eaten immediately. If they are far from the village, they usually eat the parts that are perishable, for example, the head, the marrow, and the organs (liver and heart).

Meat is not a normal part of the Rai diet and 'meat hunger' (craving for the taste of meat), is experienced. During the shepherd's three-month stay at the *lekh*, *gundruk* and *jaṅgal* greens, such as *mangan sāg* (*Cardamine hirsuta*), are the main supplements to the *chakla*. The meat brought back to the village usually consists of the thigh and leg, which have fewer calories per weight (due to the bone). Thus certain animal parts may go largely to men when they are in charge of animals at the *lekh* during the monsoon period. This is a period of 'nutritional and environmental stress,' and women and children who stay in the village may be more vulnerable to protein deficiency, due to the absence of animal products in their diet. Women are deprived of this complete protein source, containing all essential amino acids. They therefore depend upon plant protein at a time when their energy requirements are increased.

The Effect of Heavy Work Loads on Women

Shortage of time is recognized as a social disease which affects women because of traditional norms and heavy workloads. Although women's energy and nutrient needs are increased during pregnancy and lactation, they do not lower their energy output during the monsoon period when the demand for labor is greater because of the absence of the men and increased agricultural demands (Panter-Brick 1989 and Piller 1986). Piller (1986) reports that Bajracharya's 1978 survey indicated that women of reproductive age work longer hours and eat less food than adult males.

Both Rai and Sherpa women work until the onset of labor, carrying heavy loads (30-40 kg or greater) of grains, fodder and firewood. After delivery, the number of seclusion days depends on the season and work at hand, and the prescribed rest period is often adjusted to these factors. Sherpa women resume their full work duties within fifteen days post-partum. Rai women return to work one week post-partum because they experience a longer *anikāl*, which increases labor demands. One Rai woman complained of back problems after carrying approximately 32 kg of corn. Accidents occur which restrict the ability of women to work. A Rai woman fell from a tree while collecting fodder and could not work for several weeks.

Rai women have their first baby by the average age of nineteen years (range of 15-26). Although the question of age at marriage was not asked, several Rai mothers reported they were married at the age of fifteen, even though Nepalese law does not permit marriage of girls less than sixteen years of age. The risk of infant death is highest among teenage mothers (Gubhaju 1975).

Sherpa women give birth to their first baby at an older age than Rai women, an average age of 23.5 years (range of 19-30). The underlying reasons for the later age of Sherpa women is that they are often needed for farm labor and herding the high-altitude livestock. In addition to this, there are cultural factors. For example, the process of marriage requires several ritual stages and ceremonial exchanges culminating in the final ceremony referred to as *zendi* (Furer-Haimendorf 1964). Several years may have passed before the *zendi* is held. The *zendi* marks the occasion when the woman receives a share of the family's movable property, in the form of a dowry, consisting of animals, cooking pots and blankets. Weitz et al (1978) and Beall (1983) noted the importance of biological factors, because Sherpa women who experience a later age of menarchy than low-altitude farmers give birth at a later age. Beall (1983) attributed the late menarche to slower growth and delayed development, while Bangham and Sacherer (1980) remarked it may be due to the high altitude which delays the woman's fertility.

When a woman works in the fields, her infant is often left in the care of an older child or even alone. These infants are more prone to infections and nutritional stress because they may not be adequately fed and may crawl around the house and put dirt or other items in their mouths. Worm infestation is endemic among children in rural areas throughout Nepal. The mother is not able to breastfeed the infant on demand, which may reduce the hormone prolactin making it insufficient to inhibit ovulation. The mothers said the main reason for ceasing breastfeeding was pregnancy. The average length of breastfeeding is two years; therefore women's pregnancies are closely spaced, draining them of their own nutrient stores and making them more vulnerable to illnesses. Once infants are taken off the breast they are more prone to infections.

Another risk factor for women is iron deficiency anemia, which is very common among women in Nepal. Malville (1987) found 70-80 percent of the women of child-bearing age who were sampled suffer from mild to moderate anemia. Iron deficiency is caused by poor nutrition, infections and closely-spaced pregnancies which lowers labor productivity and can lead to miscarriages (UNICEF 1987). A doctor who worked at the Khunde Hospital in Khumbu for two years said Sherpa women often suffer from a state of severe depression after childbirth which may be related to iron deficiency anemia (Dawson, personal communication, 1997).

Family planning is practiced by only a few Sherpas in this area. Women receive no information about family planning and it is a 'taboo' topic of discussion. Three Sherpas who often trek and were exposed to family planning media have had vasectomies. Men often believe the procedure will make their body weak and prevent them from carrying heavy loads. One Sherpa man told me that when he is able to afford more nutritious food, such as lentils, milk and butter, he may then have the family planning procedure. Another factor is that this is a predominately agricultural area and children are valued because they are needed to farm and herd. The Rai do not practice family planning, rarely go to Kathmandu, and cannot afford to buy contraceptives.

Infant and Child Morbidity and Mortality

Poverty and malnutrition go hand in hand. According to National Nutrition Co-ordination Committee (1978) statistics, approximately sixty percent of children, less than five years of age suffer from protein-calorie malnutrition and almost ten percent of them suffer from third degree malnutrition. Thus malnutrition accounts for twenty percent of the deaths of children less than five years of age (NNCC 1978; Cassels et al 1987).

In 1987, Nepal remained one of the countries with the highest mortality rate for children less than five years of age. The mortality rate was 165 per 1,000, a decline from 297 per 1,000 in 1960. According to the UNFPA (1992), the infant mortality rate (0-1 year) is estimated at 107 per 1,000, and the maternal mortality rate, 550 per 100,000. Both Sherpa and Rai women were asked in the household survey and during the women's interviews about infant and child deaths within their families. Out of twenty Sherpa households interviewed, nearly one-half (nine) of the households reported deaths of children less than five years of age. There were a total of eighteen deaths reported, of which seven were infants. The gender of the infant was not reported in all the households and cannot be included in the mortality findings. Of the 32 Rai households surveyed, ten households reported 23 deaths and of these, ten were infants. Several families reported more than one death and one Rai woman had lost six children. The woman said she had been sick after childbirth and her breastmilk ceased after three months. Three of her babies had to be fed millet flour mixed with water as their main food source. The mothers said the main causes of death for their infants and children were: "bloody" diarrhea, headache, stomach ache, worms, measles, vomiting, fever and falling in the fire. Two twins died three days after birth because they "would not drink the mother's milk."

Both Rai and Sherpa mothers sought out traditional healers, *dhāmīs*, and *lamas*, when their babies and children were sick. The traditional healers performed rituals (*dewa pūjā* and *cintā*) and a chicken was often sacrificed (explained in Chapter 6). After the completion of the ritual, the mothers said their children either improved or died. The male community health volunteer (a Brāhman) rarely comes to the area and confessed to me a major reason is he does not like the leeches which are present six months a year during monsoon. Only one mother reported using the nearest health center, a one-day walk over very rugged terrain. For the farmers, this means two days of labor would be lost. One mother went to Khandbari, the main district center (a four-five day round trip) to get worm medicine. Another mother said she went to Froso's house (the author) for Western medicine. The women use *chiraito* (*Swertia chirata*) to treat a child who has a cold or fever. The medicinal herb is pounded in a mortar, cooked in water and given to the child as a tea-like beverage. *Mās* (*Phaseolus mungo* Roxb., black gram) are cooked up to treat a child with diarrhea. But often the disease is left to 'cure itself.'

Food Beliefs During Pregnancy and Lactation

Rai and Sherpa women do not perceive pregnancy as a period of nutritional stress, and neither group reported increasing their food intake during pregnancy. Pregnancy is said to disrupt the hot-cold balance, and foods which can restore an equilibrium are eaten. Sherpa women reported they drink two glasses of *chang* daily throughout pregnancy and *raksī* less frequently (a small glass once a week). The Rai women drink *jāḍ* 2-3 times a week.

Food beliefs can be neutral, beneficial or at times detrimental by unknowingly eliminating a needed nutrient from the diet, especially during critical periods in the life cycle. Specific food taboos for women are usually of limited duration and are largely the result of symbolic factors which may occur at nutritionally sensitive periods such as birth and lactation, when both mother and child are vulnerable (Jerome et al 1980; Rosenberg 1980).

Beliefs Regarding Hot and Cold

As mentioned in Chapter 6, food proscriptions forbidding the consumption of *jaṅgal* ingestibles emerge during stages of child-bearing. In the puerperium, food proscriptions are most marked. After childbirth, some Rai and Sherpa women experience excessive blood loss (heat) and enter a cold state. They drink only millet *chang*, *jāḍ* and *raksī*, all hot foods. An egg, if available, may be added to the *raksī* by the Sherpas. These dietary rules are said to prevent the mother from catching cold and to help replace the heat which had been lost.

After the ingestion of protein foods, the body's resting heat output is increased by twenty to thirty percent, four to six times greater than that of carbohydrates and fats (Burton 1965). Laderman (1981:490:9) asserts that this specific dynamic action of protein may be responsible for the "subjective feelings of internal heat claimed by Malays to follow meals featuring 'hot' foods." The offering of chickens, other animals and eggs during curing rituals may help to restore the balance of the ill person because the ritual offering is later consumed.

Fruits and vegetables, considered cooling foods, are proscribed because they are associated with illness and disease. An example of a negative food belief is that lactating Rai and Sherpa women, restrict *sisnu* from their post-partum diet for fifteen and five days, respectively, because they feel that it will affect the quality of their breastmilk. *Sisnu* is a 'cold' food, and considered 'too tough' for the baby. The belief is that if the mother eats it, the baby will become ill. According to Levitt (1988), it is a common belief that if cold foods (green vegetables, including *sisnu*) are eaten by the mother, they can cause cough, asthma, green stools and dysentery in the newborn. Other *jaṅgal* foods are also restricted. Gittelsohn (1991) cautioned that an avoidance of these foods can considerably reduce the intake of Vitamin A, iron and other nutrients. However, because of the short duration of the restriction period, these dietary proscription may not be a major nutritional concern.

The avoidance of certain *jaṅgal* ingestibles has an underlying rationale because they may contain toxins which can affect the newborn baby. The toxin "overdose" factor may override the nutritional benefits to the infant. Cassava contains cyanogenic glucosides which are capable of liberating large quantities of cyanide by hydrolysis (Hetzel 1990) unless evaporated out when cooked in an open kettle (Standal, personal communication, 1997).

These food beliefs affect both staples (coarse grains) and complementary foods because the texture of the food determines whether it is considered hot or cold. Corn or barley *chang* and *dhīro* are restricted because they are rough foods and millet is the preferred grain.

The preparation of the food is said to increase its "heating" effect. If maize or soybeans are roasted, they become hot foods. After greens are dried, for example *gundruk*, they are considered to be less harmful to the health of the person. The Sherpa women have said sun drying the 'wild' vegetables removes the bitter taste. Soups are boiled and *chang* is also served hot. Milk must be heated because Sherpas believe that cold milk or buttermilk can cause a stomach ache. I was told that the *dhāmī's* wife had died five years ago after giving birth because she drank cold water. This is another indication of the importance people attribute to the hot-cold theory and its influence on one's health, particularly at critical times during the life cycle.

Rai and Sherpa Birthing Practices

Childbirth for Rai women usually takes place at the household of her affinal relations. Usually the husband helps with the preparation activities prior to delivery. If the infant is a boy, a long knife (*khukuri*) is used by the mother or a close female friend to cut the umbilical cord, while if the infant is a girl, an iron sickle (*hasiyā*) is used. The iron sickle or knife is not always sterilized and bacteria can be transmitted to the infant. Neonatal tetanus is considered to be a major reason for the high infant mortality rate in Nepal (Adhikari and Krantz 1989). Although immunizing women against tetanus with the tetanus toxoid is considered the most important means of prevention, the health worker rarely comes to this remote area.

Rosenberg (1980:192) points out that, "Taboos increase the dependency of women on men, and strengthens the social stratification system. They also give men greater control over critical parts of the food supply." I disagree, however, because I found that childbirth is one life event when the unequitable distribution of protein food favors the woman. As mentioned in Chapter 6, a chicken or rooster is sacrificed after a woman gives birth. The mother is given priority for the protein foods, a chicken soup served together with the staple. *Chang* or *jãḍ* is drunk and butter is added to the mother's meals. A Tamkhu Sherpa relative gave a huge bag of butter to a poor Sherpa woman who had given birth. Kinfolk form an important supportive network during critical periods in the life cycle. Post-partum women are given butter and occasionally eggs which are rich in Vitamin A. *Caũrī ghiu* (unclarified butter), was bought from the Sherpas and given to a Rai woman for five days after childbirth in order 'to give her strength' and 'make her milk flow.' This practice is nutritionally beneficial since it provides an important source of energy needed for the increased caloric demands of lactation and nourishment after childbirth. These foods help to ensure an adequate intake of Vitamin A for the mothers because leafy green vegetables are "cold foods" proscribed during the pueriperum period.

Last, I want to mention that these hot-cold classification systems have rules that do not necessarily imply actual intake or avoidance of a particular food. Within an ethnic group, there is variability amongst its members, a result of many factors, such as socio-cultural, education, economic and geographic. Thus ethnographic observations allowed me an opportunity to observe when rules are followed or may be broken. The type of foods the daughter-in-law receives depends on the economic status of her husband's parental household and the generosity of her mother-in-law, who controls the food supply for all household members.

Infant Feeding Practices

Infant feeding practices are an important aspect of childcare. One day after birth, when a Sherpa mother's milk did not come, the baby was given butter to eat. The baby did not cry, even though she was not fed breastmilk. A 42-year-old Sherpa woman who was ill said her breasts hurt and her breastmilk would not flow. Her daughter acted as a wet nurse during the interim period, until the mother's milk began to flow. One Sherpa mother reported that she ceased breastfeeding her infant at nine months and at this time gave her infant cow's milk and an "adult" diet. Joshi (1982) compared the composition of milk of different animals, along with that of breastmilk shown in Table 35.

Table 35

Composition of milk (gm/liter) of different animals and breastmilk

Species	Sugar	Protein	Fat
Buffalo	38	62	125
Cow	45	35	40
Ewe	50	67	70
Goat	47	33	40
Nak	46	109	85
Woman	75	11	35

Source: Joshi (1982)

Human milk contains more lactose (milk sugar) and the least amount of fat compared to all the other milks shown in Table 35. Cow and goat's milk have a calorie value closest to human milk, however, because their protein content is higher, the milk would need to be diluted with boiled water so its composition would be more similar to that of mother's milk. Sometimes sugar is added to increase the simple carbohydrate level. Buffalo milk has the highest fat composition of all the animals and would be too rich for the infant if used undiluted as a substitute or complementary food for breastmilk. *Nak* (female yak) milk has the highest protein content compared with the other animals and would significantly contribute to the overall protein intake of adults and children. However, if the milk is given undiluted to an infant, the large amount of protein could cause digestibility problems and possible allergies for the infant.

Pediatricians and nutritionists assert that breastmilk is usually nutritionally adequate for the infant for the first six months of life and after this age, supplementary foods are needed. However, among the Rai and Sherpa communities, supplementary foods are given much earlier than has been recommended. The mothers said that they feel the breastmilk is not enough and need to give additional food. This increases the chance of infection and infantile diarrhea if the foods or serving utensils are contaminated.

Usually at one month of age, the Sherpa infant is given *liṭo*, a thin paste of millet flour mixed with butter and salt, and then cooked in a ladle over the open fire. I observed one mother who firstly masticated the *liṭo* mixture and then fed it to her three-month-old infant. Although the infant did not like the mixture, the mother interacted a lot with her infant during the feeding period. The mothers said that no other food is introduced until the child is one year of age and whatever the child will eat, including *chang*, *raksī* and chili are given. The cultural rules are not strictly adhered to and there is individual variation among mothers. In many households, I saw infants as young as three to six months fed potatoes, bananas, or other foods. This is supported by Pelto (1984:41) who asserts that "intra-cultural diversity in beliefs and practices are to be found even in very small, isolated communities."

Child Feeding and Snacks

The growth patterns of Nepali children in their second and third years of life are associated with malnutrition which leads to physical stunting, often combined with gastroenteritis and measles (Farquharson 1976). This age group is often left at home alone by the mother who leaves with a younger infant on her back in a basket to work in the fields all day. The child has leftover food from the morning meal to snack on while he or she waits for the mother to return home in the evening. Oftentimes the child may wander to another house, if there is a grandmother or other kin, nearby, for food handouts. In the Rai village, I noticed young children with large abdomens indicative of kwashiokor.

One Sherpa woman was pregnant and had taken her daughter off her breast when she was thirteen months of age. The mother said her child, now fifteen months old, drinks two to three bottles of *chang* daily and eats potatoes with chili. She could not give her child cow's milk because they were poor and had no milk-producing animals. A reduction in dietary diversity can create nutritional problems. For example, three times as many potatoes are required to equal the food value of the same amount of grain. Because of the high bulk of the *chang* and potatoes, the child's appetite can become quickly satiated before his or her nutritional requirements are met, thus resulting in malnutrition. The toddler's stomach capacity is small and frequent feedings are important. Adding oil or butter to the staple mixture can increase the calorie content.

The large stomachs of these Rai children are a sign of malnutrition.

Because of poverty, Rai and some Sherpa women are often not able to provide snacks for their small children. Proper snacks can be beneficial because they provide needed nutrients that are lacking in the two main meals. Leftover *chang* grains, often eaten as a snack by children, are an important source of iron and other nutrients. Young children were frequently snacking on 'wild' berries, which can boost their diet with nutrients lacking in their meals.

This chapter has shown that food is as much a social resource as it is a collection of nutrients. The way the food is produced, stored and preserved, prepared, cooked and distributed, can profoundly affect the nutritional health of a community as much as its total quantity. Rai and women's use of *jańgal* ingestibles is very important for household nutrition security. The linkages between *jańgal* ingestibles and household food security are shown in Figure 7 which has been adapted from the model used by Ogden (1990:21).

Rai and Sherpa women have developed behavioral strategies which vary according to their economic status and indigenous knowledge system of the *jańgal*. These strategies help them to alleviate seasonal and annual shortages in their food supply. Rai women use *jańgal* ingestibles as a regular supplement in their diets, while most Sherpa women gather them seasonally for variety and taste. Even when household food security needs are being met, social factors prevent an equal allocation of food within the household which can gravely affect the women.

Figure 7. Relationship of *jańgal* ingestibles to household food security.
(+ positive effects, - negative effects)

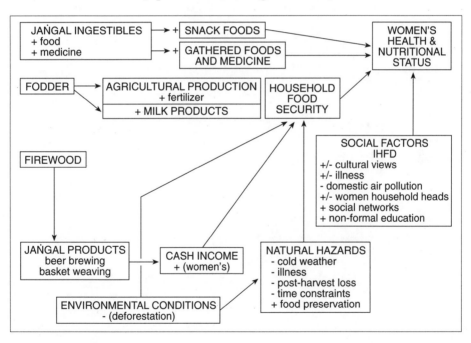

Source: Adapted from Ogden (1990:21).

CHAPTER 8

Swidden Agriculture and Trade

Farmers with too little fertilizer and insufficient permanent fields to supply food for their families are forced to resort to swidden agriculture as a means to supplement their subsistence base. A fortunate side effect of this is an increase in access to *chiraito* (*Swertia chirata*), which flourishes in fallow swidden fields for several years after the initial crop is harvested. *Chiraito* brings a good price on the market, and is the main source of cash. The trade of *jangal* resources – especially *chiraito* – at the *jangal* gate is a major means to alleviate poverty among poorer Rai and Sherpa farmers.

Case studies of *chiraito* and *lokta* (*Daphne bholua*) will be used to highlight indigenous rules regarding natural resource management and exchange relations between collectors and middlemen, and among collectors involved in the sale of these resources. They will also show that as the *jangal* becomes more essential in subsistence and/ or cash value, certain resources may be in danger of becoming scarce and so will be controlled by new local regulations.

Swidden Agriculture

Rai and Sherpa farmers practice swidden agriculture in temperate regions, and *chiraito* flourishes in these fallow swidden fields. Literature on swidden agriculture in temperate regions is sparse compared to studies of swidden agriculture in tropical regions, particularly Southeast Asia (Conklin 1954; Dove 1983, 1993). One reason is that nowadays swidden cultivation is not as extensively practiced in temperate areas (Russell 1968). For example, Ghirmire (1992) reported that more than 75 percent of the households (4,531) in the tropical Siwalik region of the Nawalparasi District relied on swidden cultivation as a major subsistence strategy. The Rais, however, practice swidden cultivation in temperate areas. In neighboring Assam and Meghalaya, swidden cultivation is also practiced by hill tribes in temperate regions (Standal, personal communication, 1997). A direct comparison with temperate regions may not be possible because of the different climatic, ecological and other factors affecting this farming technique. Nonetheless, the underlying socio-economic factors which make the farmer resort to this practice, such as land tenure patterns and the availability of land and labor, may be similar in these diverse eco-systems. Swidden agriculture yields high returns per unit of labor but is disapproved by national governments and international development agencies, "ostensibly because of its high ratio of fallow to cropping time and its purportedly destructive environmental impact and nomadic character" (Dove and Kammen 1997:93). And the colonial officials in India viewed swidden cultivation, practiced by the Baiga, as a hindrance to commercial forestry and an obstacle to their control of the forest (Elwin 1994).

Swidden Agriculture Practiced by the Rais

Swidden agriculture, a low intensity form of agriculture, is also known by various terms such as shifting cultivation, slash-and-burn agriculture, fire agriculture, and by local terms, *lose* and *khoriya*. There are numerous types of shifting cultivation described by Conklin (1954:3). The Rais fit the category of supplementary swidden farming, because swidden cultivation supplements their permanent fields, through necessity (poverty, insufficient lowland terraced grain fields and high returns for farm labor).

The Rai have historically practiced traditional swidden agriculture. The importance of swidden cultivation is exemplified by its use over a period of about sixty to seventy years ago to keep track of Rai births. Children's ages were counted by the number of years a swidden field had been left fallow during that time period, usually in nine to ten year cycles.

Nowadays swidden agriculture is extensively practiced by the Rais because of the diminished size of their permanent field holdings. Conlin and Falk (1979) reported that for 75 percent of the population in the Eastern Hills, the average household land holding is less than 0.5 hectares. The amount of land is continuously diminishing because of the tradition of dividing land equally among sons and selling land to obtain cash needed for funeral expenses. In addition to this, crop yields have been decreasing and the farmers say this is because the soil is not as fertile as before. Khadka (1997) suggested the following as possible factors accounting for the soil infertility: a lack of manure as the number of livestock are decreasing, a lack of compost because the children are going to school or doing other chores instead of collecting foliage from the forest, erosion of top soil because of increased cropping intensity, and alternatively, crop losses due to diseases and pests.

Rais are often seen clearing *jaṅgal* land, and swidden fields are common around Yangden. The Rais are more dependent on swidden agriculture than the Sherpas, who are mainly agropastoralists and dependent on their livestock for plowing, manure, the consumption of milk products and the sale of animals, butter and wool. Although swidden agriculture is practiced by Sherpas, animal husbandry is more important as a supplement to fixed field agriculture.

Methods and Benefits of Swidden Agriculture

In swidden agriculture, vegetation is cut and subject to controlled burning. Whole trees and tree stumps are often left standing in the cleared field to avoid unnecessary disturbance to the top soil and promote regrowth of the forest by preserving the original root system. Another advantage to leaving the trees standing its that it requires far less labor, an important energy limiting factor in areas where labor is scarce (see Dove 1985). Khadka (1997) suggested the stumps may be left because the root system of the undisturbed stumps hold the soil firmly and prevent erosion during the monsoon when the rainfall intensity is quite high. The monsoon coincides with the major crop growing season of maize and millet. This method was also practiced in the Ozarks and Appalachia in the United States, an important adaptive strategy where weather is unpredictable and fluctuations in temperature occur (Otto and Anderson 1982:141).

Although deadenings struck outside observers as being strange and primitive, the partially cleared fields offered highlands farmers a means of coping with such agricultural limitations as unseasonable frosts, steep slopes, and thin, leached soils. The air currents generated between a deadening and the neighboring woods fostered the formation of dews and fogs, thus often saving corn crops from untimely frosts. Deadenings also had the effects of retarding surface winds, reducing surface evaporation, and retaining moisture-laden snow packs during the winter. In turn, the root networks of the girdled, dead trees helped retain the thin topsoil for a few seasons of cultivation, even on the steep slopes. And the troublesome undergrowth ultimately proved to be beneficial, for when farmers fired the piles of brush which they had grubbed up, the burning released fertilizing ash, killed insect pests, and baked the soil so it could be more easily worked (Otto and Anderson 1982:141).

The farmers say that large *utis* (*Alnus nepalensis*) trees are not often cut because men are temporarily absent due to migration and cutting large trees is considered "men's work." An additional reason is that if the tree falls down in the swidden field, it covers the place where crops are to be planted. Khadka (1997) reported that in other areas of Nepal, *utis* leaves are considered a good source of crop nutrients because they decompose quickly. *Utis* roots fix atmospheric nitrogen (a much needed nutrient) which the companion crops also utilize. Russell (1968) noted that tree stumps serve a dual purpose: they provide shade which suppresses the growth of weeds and new growth directly sprouts from the stumps. Therefore there may be several possible reasons for not cutting *utis*.

The farmers said that after the vegetation is burned, the "nutritious" ashes are used as compost and natural fertilizer for planting crops. The ash fertilizer is of vital importance for these farmers, especially the Rais, who lack livestock. Another important factor is that the land is dug by hand and not plowed, avoiding the cost of hiring oxen (Humphrey 1980). Swidden farmers also exploit a broad range of 'wild' resources in the fields (Strickland 1990). Meilleur (1994:262) points out that investigators have noted that swidden agriculture is "a complex multiyear strategy of land management, often incorporating 'wild' gathered species into its later successional stages."

Crops Grown on Swidden Fields

Swidden agriculture acts as a supplement to fixed field agriculture and relieves the farmers from food deficiency. Swidden farming systems usually have a diverse range of crops planted together on a single plot (Strickland 1990). In Nepal, maize, millet, soybeans and rice beans or *mās* (*Phaseolus calcaratus Roxb.*) are planted in the burned fields for one or two seasons, depending on the availability of household members to weed the fields. The combination of maize and millet has the advantage of providing a mixture of grains of superior nutritive value (Daniggelis 1994b). Khadka (1997) described other advantages of intercropping. Maize and millet complement each other during the growth period, having different maturation periods. Maize has deeper root systems than millet; hence they extract nutrients from a different soil depth. These crops are symbiotic because there is less competition for nutrients, millet is more adapted to and efficient under shade. These two crops together combine to improve ground cover (due to higher plant densities) and better protect the soil from erosion than maize by itself. From the same piece of land, the two crops satisfy the diverse needs of the family and ensure food security. And even if one crop fails, the other crop may be able to help compensate for the loss. In the harsh conditions of the hills, subsistent farmers need to spread their risks by practicing multiple cropping. Maize is the major staple; however, millet is also an important crop. More than 75 percent of millet (in the middle hills) is turned into alcohol which is necessary for many rituals, funerals, fairs, *pūjās*, festivals and to serve guests. Millet also brings a higher price than maize, so it is a good source of income both from grain and alcohol sales. Both crops combine to provide fodder for livestock during the lean winter months.

After these crops are harvested, wheat, a major winter crop, is planted along with barley. The increased demand for human labor in weeding, rather than the decline of soil fertility, is given as the reason for abandoning the swidden field. Freeman discussed that among the Iban in Sarawak, the size of the field cut is dependent on the availability of family members to weed the swidden fields (see Freeman 1955, pp 90-7). Boserup (1965) first postulated that population growth leads to an intensification of agriculture. Dove (1985) asserts that studies of agricultural development must also address state policy and interests which differ from how the swidden cultivators view their landscape.

Proposed "Regulation" of Swidden Cultivation

Two types of swidden agriculture, *lose* and *khoriya*, are referred to by the Makalu-Barun Conservation Project (MBCP). *Lose* is a form of rotational agroforestry whereby naturally regenerating trees are harvested for firewood, the remaining brush and weeds are burnt, and maize planted. If the rainfall is high, the seeds and rhizomes are widely distributed, and soil cover is rapidly regained. This agroforestry practice limits erosion to levels estimated to be lower than permanent upland fields, permits nutrient recycling, and produces significant quantities of foodgrain and fuelwood as well as miscellaneous non-timber forest products (Shrestha et al 1990). *Lose* is recognized as a legitimate agroforestry technique by the MBCP. Shrestha et al (1990) found that in almost half of the households surveyed in the Makalu-Barun National Park and Conservation Area, *lose* makes up about 55 percent of their agricultural landholdings.

On the other hand, *khoriya* is swidden which expands into previously uncultivated forest land. The MBCP asserts the practice of *khoriya* is destructive to the environment and plans to regulate it by preventing its expansion into "mature pristine" forests. The MBCP will work together with forestry user groups to provide perennial alternatives (Shrestha et al 1990). According to Shrestha (1989), *khoriya* is most detrimental to forest ecology and contributes to the total extinction of a large number of biological species. Although it is perceived that the practice of *khoriya* degrades the environment and leads to species extinction, the author points out that the ecological, as well as the economic impact of *khoriya* in the Arun Basin has yet to be studied systematically. Shrestha is making conclusions about *khoriya* that he even admits are unproven.

Swidden agriculture, practiced at appropriate levels of population density, causes little long-term degradation of the productive capacity of the environment (Rambo 1980), when the fallow period is long enough for the soil to recover fertility. Otto and Anderson (1982) assert that true forest fallowing, Highlands South land rotation used in the Ozarks and Appalachia, required twenty to twenty-five years until the land was reforested. Among the Rais and Sherpas, *lose* fallow cycles range from five to ten years. The fallow cycle varies with pressure on land resources and also available household labor. Kleinman et al (1995) asserted that slash-and-burn agriculture with a ten to twenty year fallow period can provide effective long-term control of erosion and conserve nutrients even on relatively steep slopes of up to twenty percent. The authors caution that periods of one to five years are too short for "sound" slash-and-burn agriculture. For the Baiga of India, six to seven years is sufficient time for the fallowed *bewur* to be re-cultivated (Elwin 1994).

The regulation of swidden areas within the Makalu-Barun National Park and Conservation Area, as well as throughout the Sankhuwasabha District, will greatly reduce the availability of *chiraito*. This regulation can lead to the loss of an important source of cash income for the rural farmer. Dunsmore (1988) warned that a much more vital and sensitive situation will arise with regard to swiddens cut on areas which are legally government land but in practice have continued to be used by households with traditional rights to them. The Rais previously had unrestricted access to *jangal* resources through community-owned (*kipat*) land. These land tenure rights are very critical for these farmers who own little cultivated land. According to customary law, the farmer who first cleared "virgin jungle" stakes his claim to it and passes it to his patrilineal descendants who would have permanent use rights to hold the land as *kipat* (Caplan 1970). Between 1888 and 1957, there were no laws regulating the rights of tenants cultivating *kipat* lands in Eastern Nepal (Regmi 1978). The *kipat* system gave the Rais power over the *jangal* resources. In swidden systems, a common error of policymakers has been to assume that the fallow land is unused (and unowned), that swidden farmers use communal land and that these systems are inefficient (Dove 1983).

Trade of *Jangal* Resources

Chiraito (Swertia Chirata)

Chiraito (*Swertia chirata*), a 'wild' indigenous plant, is a perennial herb. The Sherpas refer to *chiraito* as *tikta* and the Kulung Rai call it *khipli*. The two species of *chiraito* which grow in this area are *dankhle* and *pāte*. *Dankhle* is a tall plant (may grow up to two meters) with a woody stem and is found in greater abundance than *pāte*. *Pāte*, is smaller than *dankhle*, more leafy (has less weight per biomass), and grows in fewer places. The *chiraito* merchant prefers *dankhle* and pays double the price paid for *pāte*. According to Khadka et al (1994), the reason for the difference in market value of the two species is that *dankhle* has medicinal properties, while *pāte* is used only in dye preparation.

Chiraito grows between 1,800 and 2,500 meters elevation in sunny areas with little tree cover. For this reason, it flourishes in fallow swidden fields. This is a good example of co-evolution of swidden *jangal* and a 'wild," but economic plant.[37] *Chiraito* appears the year following the harvesting of maize and millet and has a life cycle of about four years. According to the farmers, the amount of *chiraito* varies with the extent of swidden agriculture.

Chiraito collection is labor intensive because of its dispersed distribution. A Rai man received permission from a large landowner from Dobatak to collect *chiraito* on his land because the Rai holdings were small. It took six days for him and his wife to collect five kg of *chiraito*. *Chiraito* collection is not differentiated by gender or social

---

37 I want to acknowledge Michael Dove for this important observation.

groups. Even children, while walking to and from the school in Yangden or herding, collect *chiraito*. Although the collection of *chiraito* coincides with the busy harvest season, it often complements other farm activities (herding, collecting fodder and fuelwood) where labor needs can be met. Another advantage of *chiraito* collection is that it helps maintain a household's labor unit and prevent out-migration for wage jobs.

Chiraito is picked from August to November, mainly on the farmer's own land, but occasionally from a friend's private *lose* or from the "*jangal*" (which the local people view as communal land). The farmers collect the entire plant and say the seeds disperse naturally during collection, which enables subsequent regeneration. Khadka et al (1994) assert the practice of cutting *chiraito* at the ground level ensures good regeneration.

Jangal resources also vary according to whether they are ready for final consumption immediately after harvesting, or are ready with only minimal processing, such as drying *chiraito* (*Swertia chirata*) and *juniper* (*Juniperus indica*). The *chiraito* is first tied into bundles and left to sun dry near the farmer's houses. Later it is stored under the houses until the merchant arrives. Other resources, such as various types of bamboo, entail the labor intensive task of making them into baskets or other implements before becoming useful to consumers.

If cash is needed quickly for salt buying trips, weddings, funerals or school fees, *chiraito* is picked early during the monsoon (June to August) on common land. Early harvesting of *chiraito* occurs when the plant is green and still flowering. The seeds have not matured, so the cycle of natural propagation is prevented. The villagers also receive less money from the traders for the immature plants (Daniggelis 1994a). Brokers and traders prefer buying well-matured plants because the weight loss during storage is less than one-half that of immature plants (Khadka et al 1994).

My research area is remote and far from Hile, a trade center and roadhead four to five days walk to the south. Thus the farmers must depend on middlemen to buy their *chiraito*. The merchants, usually Rais, come from the neighboring villages of Chhoyeng, Pelangma, Yachamkha and Yaphu (see Figure 2, Chapter 3) to inform the Sherpas and Rais of the current market price for *chiraito* in Hile. The merchants often give an advance payment about two months prior to collection of the *chiraito*, usually to those farmers who had harvested a lot of *chiraito* the year before. For example, in Dobatak, the two households which sold the most *chiraito* in 1992 were given Rs 1,200 and Rs 1,500 advance payment, while all other households received only about Rs 100.

Often another *chiraito* merchant may follow the first merchant and offer more money for the *chiraito*. However, the first merchant assures the people not to worry because he will meet the higher price of any merchant who outbids him. From October to December, the *chiraito* merchants come to the research area to purchase the *chiraito*. In November, I saw *chiraito* drying outside many houses in Yangden. The *chiraito* is weighed, and the money is paid.

Chiraito is heavily traded from the hills in East Nepal via the Hile-Basantpur roadhead. During 1991-92, *chiraito* comprised about 75 percent of the total cash value and 60 percent of the total volume of trade occurring in Hile and Basantpur (Edwards 1993). Khadka et al (1994) assert that the export quantity of *chiraito* is declining due to improper management and over-exploitation.

A porter carries chiraito to the roadhead in Hile

The transnational trade with India in medicinal plants produces a huge revenue. The Foreign Trade Statistics for Nepal reported that *chiraito* comprised 32 percent of the export earnings in medicinal herbs to India for the period of 1990-91. These figures undoubtedly are an underestimate due to the existence of an illegal trade in medicinal plants which exists between Nepal and India (Rawal 1994).

Chiraito is transported as raw material to India for the drug *chiraita*. *Chiraito* contains the chemical compounds chiratin and ophelic acid used as the bitter principle and febrifuge in ayurvedic medicine (Burbage 1981). Chiratin is also used for making Quinone, a malarial drug (Khadka et al 1994). The Rais and Sherpas use *chiraito* as a treatment for colds, coughs, fever, stomach ailments, sore throat, "bloody" diarrhea and fever. The entire dry plant (including roots, leaves and stems) is beaten into small pieces and soaked overnight. The liquid is then consumed by the ill person.

The *Chiraito* Survey

In February 1994, a semi-structured survey was carried out among all Rai and Sherpa households in Gongtala, Dobatak and Yangden, to determine the extent and importance of *chiraito* collection. See Chapter 3, Figure 2, for location of the *chiraito* study area. Another objective of the survey was to assess the potential for cultivation and processing of *chiraito* at the village level. Saisima was not surveyed because there is no *chiraito* growing nearby. Swidden agriculture is not practiced because the people are Buddhists and the "fire would kill all living things, including insects."

Table 36

Amount of *chiraito* collected and sold in Gongtala, Dobatak and Yangden (1992-93)

	Gongtala 13 HHs	Dobatak 4 HHs	Yangden 32 HHs
1992			
Number of households collecting *chiraito*	3	3	8
Total number of kg	12.5	115	91.3
Ave. kg per household	4.3	38	11.4
Rs/kg (range)	40-54	24-48	20-48
Total *rupiyā* (households)	505	4,720	3,475
Average Rs/household	172	1,573	435
1993			
Number of households collecting *chiraito*	6	3	17
Total number of kg	65	438	290
Ave. kg per households	11	145	17.5
Rs/kg (range)	60-64	40-64	40-64
Total *rupiyā* (households)	3,970	26,800	17,215
Average Rs/household	662	8,933	1,013

Note: 42 *rupiyā* = US$1 at the time of the study. Rate per kg depends on merchant's quoted price. HHs = households.

Table 36 shows that the largest amount of *chiraito* is sold by only three households in Dobatak. In Dobatak, a lot of *chiraito* is available because land holdings are large and swidden agriculture is extensively practiced. In all the villages, the number of kg of *chiraito* sold has more than tripled between 1992 and 1993. The price for one kg of *chiraito* has also increased. This situation is analogous to Dove's (1993) discussion of rubber cultivation and swidden rice in Borneo whereby the two systems result in not only minimal competition in the use of land and labor, but promote mutual enhancement. Rubber cultivation meets the need for market goods and swiddens provide for household subsistence requirements. Dove (1993b:136) asserts that "the special virtues of such composite systems merit greater attention by development planners."

The farmers have been selling *chiraito* for an average of 2.5 years (range 1-10). In 1993, the income from *chiraito* sales was used for: foodstuffs, including salt (16 HHs), clothes (10 HHs), to pay off loans (3 HHs-Yangden) and to buy two cows and a pair of ox (the Dobatak HH which earned Rs 16,000). *Chiraito* collection is a supplement to other cash income sources, trekking (Sherpas), selling livestock and farm wage labor (Rais). A 12-year-old Sherpa boy who collected 2.5 kg of *chiraito* while herding bought a pair of shoes with the money he had earned. The timing of the *chiraito* sales coincides with the festival of *Dasaĩ,* which helps supplement expenses incurred during this period.

Lokta (Paper Bark Bush)

Lokta (*Daphne bholua* Buch-Ham., ex D. Don), a paper bark bush, grows as an understory shrub between 1,500 to 3,000 m in coniferous and broad-leafed temperate zone forests. The one- to three-meter plant is found in the *jaṅgal*, and may be "coppiced" just above the ground, a technique which allows for regeneration. *Lokta* may be promoted by partial disturbances of temperate forests of the Arun Basin. The inner bark is dried and generally sold directly to small papermaking enterprises. This provides a local source of income.

Two types of *lokta* are collected by Rai and Sherpa farmers. The *lokta* growing near Yangden, called *aul ko lokta*, sells for Rs 12 per *dhārni*. The other variety, referred as *kāri lokta*, is found at a higher elevation (above Gongtala and Saisima). The farmers receive a higher price for the latter variety, about Rs 25 per half *dhārni*. Several farmers remarked that the collection of *lokta* is difficult and the money received is very little. Usually Rais or Tamangs come from neighboring villages and inform the local Sherpas or Rais of the current market price for those products which are in demand in Hile. If it seems profitable, during the agricultural slack period, a family member, whoever is free, will cut the *lokta*. In Yangden, *lokta* is seen drying on bamboo poles in the sun, which saves firewood. They are extracted in September or between March and June and later traded. Khatri (1994) noted that expanding *lokta* cottage industries provide direct employment for 1,500 families and that 100,000 tons were produced in 1984. In 1984, a national study by the Department of Forests revealed that Sankhuwasabha District provided 14.1 percent of the total gross stock of *lokta* for Nepal (Bhadra et al 1991).

Lokta drying on bamboo poles in Yangden.

Conflict Mediation Case Studies

Problems occasionally occur in the *chiraito* and *lokta* trade. These conflicts are often mediated through informal user groups. In the short time span of only one year (1992-93), a greater number of households became involved in the trading of *chiraito* (twice as many households in both Gongtala and Yangden). An increase in household income from *chiraito* sales has become an incentive for careful management and private control. In 1993, the informal user group for Gongtala,

Dobatak and Yangden met and made rules on how to manage the collection of *chiraito*. Ownership rights over *chiraito* changed from free collection on both private and communal land to collection only on communal land (and one's own private land). The following incident, described by one of the farmers, illustrates how the new rules regarding *chiraito* collection were broken and enforced by the informal-user group.

> *This is the first time an incident regarding chiraito occurred. Previously chiraito was cheap. Only one or two years ago, anyone could come and collect chiraito from our land because it was "free." Now the price of chiraito has increased, selling for about two hundred rupees per dhārni (2.5 kg). This year there was a meeting called in Gongtala and Dobatak regarding chiraito collection. It was decided that chiraito could only be picked on one's own land. If someone picks chiraito from another person's field, a fine will be given. One day a villager saw three dhārni of chiraito hidden by a rock between the two villages. This chiraito was picked illegally. Another meeting was called to discuss the incident. The person who had illegally picked the chiraito was fined five hundred rupees.*

This anecdote showed that as the *jaṅgal* becomes more essential for subsistence and/or cash value, its resources will be controlled by new local regulations. *Jaṅgal* resources from public lands were traditionally regarded as a "free good" in which no investment was necessary. Land tenure rights to *chiraito* collection changed in anticipation of the Cadastral Survey which was scheduled to begin in 1994 and in response to higher prices for *chiraito*. But will local management be sufficient to prevent the potential problem of "scramble" described by Hardin (1968) in *The Tragedy of the Commons*? Khadka et al (1994:8) reported that problems of theft from private forest, grazing areas and cultivated areas is increasing because of the high market price *chiraito* brings.

Lower level middlemen, it is claimed, often take advantage of the 'ignorance' in marketing of farmers and gatherers. The local producer is often at the "mercy" of middlemen as to the availability and accuracy of information regarding a 'fair price' for their product, especially when they live in remote areas far from the major trading roadhead. Lintu asserts an additional weakness is that single gatherers lack organization on an individual basis and are in a much weaker position in negotiating with middlemen than a cooperative would be (Lintu 1995b). This assertion is proved false by the following anecdote describing the negotiations between a *lokta* merchant and Sherpa villagers.

> *In the autumn of 1991, a lokta merchant visited Gongtala because he planned to take the lokta, which the villagers had previously gathered and dried, on credit. He said that he would pay the villagers in cash at a later date. The merchant needed a porter to transport the lokta to a factory, a two-day walk, but the porterage rate that he offered was very low, Rs 25 per day and no food. The Sherpa men were very disturbed with the shrewd merchant's*

propositions and held a meeting. They reached a group consensus and unanimously declined the merchant's terms. The merchant appealed to his mit,[38] the oldest son of the most senior Sherpa in the village, to exert his influence and negotiate on his behalf. The merchant's mit was in a dilemma because roughly one-half of the group were his kinsmen and he had affinal relations with the others. Therefore he felt powerless to support the merchant during the negotiation process. The final outcome was that the merchant had to pay the porter a higher salary, Rs 30 per day, and provide food and alcohol. In addition to this, he had to pay the villagers in cash for the lokta prior to transporting it to the factory. The merchant then borrowed an equivalent amount of money from his mit to pay for these transactions.

Humphrey (1992) warns that cheating may occur in trade transactions. The middleman takes a risk in trading with villagers, a reason why an advance on the merchandise is only practiced on a regular basis with known people. But even then, problems may arise. Products which appear at first to be alright turn out later to be intentionally adulterated. The only sanction against this behavior may be grumbling, negative gossip and the refusal to buy next time. The following anecdote shows how a *chiraito* middleman was cheated by a cunning villager.

One Rai man from Yangden received an advance payment of two thousand rupees from a middleman for an equivalent amount of chiraito to be collected at the rate of Rs 72 per kilogram. The Rai carried 39.5 kg of chiraito to Gongtala from Yangden and stored it at a Sherpa household prior to the arrival of the chiraito merchant. When the merchant was weighing the chiraito, he noticed that twigs resembling chiraito had been inconspicuously hidden in the middle of the bundle of chiraito, mature and immature plants were harvested together and the batch was of inferior and inconsistent quality. The amount of chiraito was only 20 kg, shorting the merchant 19.5 kg He then went to Yangden to tell the Rai he had been cheated. The Rai refused to come to Gongtala to look at the chiraito and said he had met the terms of the agreement. The chiraito merchant knew he could do nothing and said he wouldn't do business with the man again.

Long standing patron-client relationships between the collectors and traders have been giving way to short-term competitively established relationships of expediency, which is what occurred in this instance.

Chiraito Processing Center

The other goal of the survey was to look at the feasibility of establishing a *chiraito* cooperative enterprise run by and for the people in order to enhance rural income. User groups throughout Nepal have become involved in processing aromatic and medicinal plants as a means of poverty alleviation (Rawal 1994). The Makalu-

---

38 In Nepal, a *mit* relationship is a social institution where bonds are established between two individuals often for purposes of providing mutual aid during periods of stress. Traders are known to choose a *mit* in a village where they plan to conduct business.

Barun Conservation Project (MBCP) is looking into the prospect of establishing these village-based *chiraito* enterprises because there is not a processing center for *chiraito* in the Sankhuwasabha District (N. Joshi, personal communication, 1993). The rural farmers who were surveyed perceive that the benefits will not be distributed equally from these *chiraito*-based enterprises. The collector often receives only five to ten percent of the wholesale price (HMG/Nepal 1988). During a visit by Khadka et al (1994) to Basantapur (a major roadhead in Eastern Nepal), a kg of *chiraito* sold for Rs 80, almost double what the farmers were paid in my area, according to Table 36. The subsistence hill farmers need to get a fair price for the *chiraito*; however, Lintu (1995b) asserts they have "little power" in the market place. According to Hammett (1994), improving the farmer's bargaining power through increased market information can enhance their income.

The remoteness of this area may be a major constraint to receiving market information. These marginalized farmers have a wealth of indigenous knowledge regarding the *chiraito* species, location, its uses and growing conditions. These farmers, however, need to have technical input as to the appropriate methods of *chiraito* cultivation and harvesting to ensure that they are ecologically sustainable. In the late 1970's, the Khandbari Agricultural Development Office gave farmers seed for the cultivation of *chiraito*. They did not know how to look after the seeds and the plants perished in the gardens (Humphrey 1980).

Cross-visitation by farmers who are already cultivating *chiraito* in other areas of Eastern Nepal and other two-way communication methods would enhance the development of village-based *chiraito* enterprises. In Ilam, Eastern Nepal, *chiraito* is presently being planted on experimental farms. Farmers from the Apsuwa Valley could be taken to Ilam and shown by the Ilam farmers how the *chiraito* trial works. Farmers teaching other farmers has proven to be very successful in other developing countries, and is currently taking place in Nepal as well. Cultivation trials of *chiraito* could be started at Pakribas Agriculture Centre which is within the Sankhuwasabha District (Burbage 1981), together with their monitoring and evaluation. The collection and processing of *chiraito* has great potential as a means of rural income generation for the rural hill farmer, but careful planning together with the local farmers must be done first.

CHAPTER 9

The Sociology of *Jaṅgal* Resources

Natural Resource Management System

I am adapting several concepts to fit the analysis of the natural resource management system that exists among the foraging farmers. The management of natural resources at different organizational levels is a concept adapted from Burton et al (1986). The authors of this volume assert that only by studying the entire natural resource management system can policy solutions be identified which may entail changes in the behavior of resource managers, as well as that of resource-dependent communities. The term 'sociology' is attached to natural resource management because it refers to analyzing the perception, use and management of natural resources by human societies (Dove and Carpenter 1992:1). I use the same term but apply it in a broader perspective to encompass the various levels of natural resource management by local and external actors.

1) The lowest level of this natural resource management system is the natural environment with the *jaṅgal* resources which the Rai and Sherpa farmers find useful. This level has already been discussed in Chapter 5.

2) Within the second level of the associated social system is the village with its political and religious decision-making organizations affecting the use and management of these common property resources. Also found here is the knowledge of the resources which affects how they are used in nutritional, religious, socio-cultural, and economic ways.

3) At the third level of the social system are found the outside actors, that is, the decision-makers at the level of national government officials, national forest administrators, development officials and scientists, who often are key players in regulating resource management. These external actors have their own cultural system which can either positively or negatively influence natural resource management.

Common Management of the *Jangal*

I will not be evaluating the sustainability of the *jangal* according to the perceptions of Rai and Sherpa farmers, based on their local knowledge systems and beliefs, because of a lack of supporting quantitative data. Godoy et al (1995:34) emphasized the importance of measuring sustainability directly, which requires "monitoring the population of animals and plants in the forests over time and the ratio of relative abundance." On the contrary, the farmer's images of the landscape will be used to create ethnic-specific cognized maps. I used these maps to help determine what the farmers pay attention to in the *jangal*. Do they perceive that there is a problem with the *jangal*? Do they presently use conservation methods to protect the *jangal* resources or are they planning to take steps towards conserving them in the near future?

I am using Bjonness's (1986:279) definition of perception as "an all-encompassing concept referring to images, memories, preferences and attitudes" which affect human behavior. Rai and Sherpa farmers were asked to discuss the condition of the *jangal* and whether any changes had occurred, and if so, what types of changes may have occurred. These scenarios were used to develop ethnic-specific cognitive maps of the environment surrounding the farmer's villages.

Sherpa Cognitive Environmental Maps

Several Sherpas who had large landholdings and had lived in Dobatak (fifty to sixty years) and Gongtala (thirty years) talked about the state of the *jangal*. They told me:

> *There is no difference in the condition of the jangal today compared to the past. More plants have grown than before. We collect wood, fodder grass and vegetables, such as sisnu and bhurmang (a 'wild' leafy green). Wood is not disappearing. All types of wood are available in sufficient quantities and several types of wood: utis (Alnus nepalensis), kapling (S) and takpa (S) (Betula utilis), are increasing in quantity. We do have to walk farther to collect it. There is no problem with the jangal because it has not changed.*

The greater abundance of plants can be linked with secondary forest growth which is often stimulated by fires. Herders set small patches of *jangal* on fire during the early spring because it will make the area more suitable for livestock grazing. The open areas produced by the fire provides economic benefits for the people because bamboo (used for woven mats) flourishes and *chiraito*, a medicinal plant discussed in Chapter 8, are important for the household economy.

A 45-year-old Sherpa woman born in Gongtala had a different cognitive model of the *jaṅgal*. Although she received some land from her parents and had bought land, her land holdings are insufficient to meet household food needs.

> *The condition of the jaṅgal is different now than when I was a small girl. The population has increased and men have been cutting a greater number of trees. Cows graze on the newly grown plants before they have fully matured. I cannot see tigers and bears these days. The jaṅgal's resources have not changed in the forest but the jaṅgal has decreased in size.*

In her childhood, Gongtala had only three houses. By 1993, there were fourteen houses in the village, which accounts for the population increase and pressure on the *jaṅgal*. She relies heavily on the supplement of 'wild' greens, mushrooms and bamboo shoots for her family's meals. Her husband (from outside the area) must trek on a seasonal basis in order to make up for the deficiency in their yearly household food stocks.

The *gönpa* in the village of Saisima is a unique element influencing the environment, because the principles of Buddhist religion are strictly adhered to by the villagers. This is apparent by the lush environment surrounding the village where swidden cultivation is not practiced. The head *lama* at the *gönpa* in Saisima discussed how Buddhism affects their subsistence activities.

> *The jaṅgal has not changed and is the same as many years ago when we migrated from Solu Salleri in 1974. Khoriya is not practiced. We believe if we make a khoriya many insects will die. We will have sinned, which will affect our reincarnation. For this reason there is a lot of jaṅgal.*

Denchen Chöling Gönpa and sacred forest in Saisima.

Rai Cognitive Environmental Maps

The Rai farmers who are more dependent than most Sherpas on the *jaṅgal* resources for their subsistence and livelihood had a perspective on their landscape more similar to the poor Sherpa woman. An 82-year-old Rai, who is the oldest man in the research area, recounted the historical changes which have occurred to the *jaṅgal*.

> *Our ancestors (during my paternal great grandfather's time) migrated to Yangden from Mahakulung six generations ago (about 160 years). At that time, the jaṅgal was a thick forest and there were a lot of bears and tigers. About three generations ago, the people began to cut the trees and much of the jaṅgal was destroyed. The people are using a lot of firewood. Compared to when I was a small boy, I think the jaṅgal is smaller. We have to collect very small branches because there are no large trees near the village. I think the farmers should plant more trees around their homes. It is important to hold a meeting to save the jaṅgal.*

Several other Rais recalled similar perceptions of the *jaṅgal*.

> *The jaṅgal is changing a lot and has decreased in size. For our families, the change is not good. Māliṅgo is scarce. 'Wild' green vegetables, bamboo shoots and mushrooms are rare and we must go very far to collect these foods. The population is increasing. The land is not enough for our families and even this land can be washed away during the monsoon. When we build a house, we need poles, but we can use the jhinguni (Eurya cerasifolia) tree on our land at the border of the field terraces. We need the jaṅgal to collect firewood so we can cook our food. We have to save the jaṅgal and call a meeting so all of us can decide on the best plan of action. If everyone would agree, each year we will change the cutting area for firewood so the jaṅgal will grow again.*

The Rai farmers are aware of the environmental situation because they have noticed changes in the type and quantity of *jaṅgal* resources. Yangden is situated in a valley near the Apsuwa *Kholā* (1,100 meters at the lowest elevation). The Rais must walk further to reach the *jaṅgal* because it is located at a higher elevation closer to the Sherpa villages.

These historical *jaṅgal* descriptions show that the cognitive models (perspectives) of the environment held by Rai and Sherpa farmers are diverse. Jamieson (1984:10) suggested that differences in ethnicity or mode of production may produce conflicts in environmental perceptions. The Rai and Sherpa's perceptions are partial impressions of the condition of the *jaṅgal*, and indicate that variability (both inter-ethnic and intra-ethnic) exists. The same area of *jaṅgal* may be used for different purposes and its resources show seasonal variability. Possible factors influencing this are land ownership, the potential for off-farm labor opportunities and economic status. These maps can be used as a catalyst to facilitate local participation in co-managing the *jaṅgal's* natural resources.

Milton (1996) points out that those local people who are most dependent on the environment are more apt to protect it. The Rais, and poorer Sherpas, are very concerned about the sustained viability of the *jaṅgal* because they are dependent on the land to meet all of their subsistence needs. Even though swidden cultivation helps to relieve food scarcity (*anikāl*) and can be intensified, the soil fertility is reduced if fallow periods are shortened and new areas of land must be used, either private land holdings or *jaṅgal* land. These cognitive maps of the environment "serve as mental templates for action and reasoning which is why they are relevant to resource allocation and management" (Nazarea-Sandoval 1995:103). Fox (1995) found perceived environmental risk to be an important factor in the design and adoption of forest policies among farmers in the central Nepal hills. The Rais and poorer Sherpas are conservation managers because they are concerned about the state of the *jaṅgal* and want to call a meeting to mobilize social concerns into a plan of action.

Individual Management of the *Jaṅgal*

A change in the environment will elicit different behavioral responses from those dependent farmers living on its fringes. Many factors can affect the different repercussions felt by the farmers. Dry wood, fodder and *sisnu* are freely collected on private land and only "green" tree felling is restricted. However, an individual acting alone deviates from the norm which is shown by the following incidents.

> In February 1992, a Sherpa felled a tree on his own land and left it with the intention of returning to chop it up as firewood. Another Sherpa came and took this wood to build a chicken house without permission. The Sherpa whose tree was stolen was annoyed by the incident, however, he took no action. The Sherpa who had already swiped the tree cut a tree on another Sherpa's private land. The Sherpa whose tree had been felled decided not to confront the culprit about it and this dispute was not taken to the village court.

The culprit was married to a poor sick Sherpa woman who was respected in the village (her father was one of the first inhabitants). He had little land and wanted to supply his household with sufficient firewood before he went off trekking for several months. Historically land disputes (access rights and boundaries) have been a common occurrence in Nepal (see Forbes 1993) and are a pressing issue of concern for the farmers because of the anticipated Cadastral Survey.

Cultural adaptation to the environmental changes of the *jaṅgal* and its resources is apparent on the part of the Rai and Sherpa farmers who are employing various individual management strategies. The farmers told me they are planting fodder seedlings on the banks of their private land because in many areas of the *jaṅgal*, fodder is not as abundant as before and planting it is a labor saving technique. The Sherpa farmers have transplanted *nibhara* (*Ficus roxburghii*) and *dudula* (*Ficus nemoralis*) and Rai farmers, *utis* (*Alnus nepalensis*) and *chap* (*Vimalis*) on their land.

Utis was chosen because it grows the quickest and tallest of all the fodder species. Carpenter and Zomer (1996) also observed the semi-domestication of fodder trees in the Makalu-Barun National Park and Conservation area.

Fox (1995:37) noted in a village study (1980-1990) of forest use practices in the central Nepal hills that the major cause of forest degradation had been grazing and fodder collection for the large number of livestock. In 1990, the forests were in "much better condition" compared to the period when the researcher had first conducted his assessment in 1980. Fox attributed this to several factors: an increase in the planting of fodder trees on farmer's private land, the introduction of a new tenure regime for forests and the assistance of non-governmental organizations. Gilmour (1988:344), found that for at least two decades farmers in two districts of Central Nepal have been increasing tree cover on private land. The author cautions that the deforestation controversy has a broader definition which needs to consider the "whole landscape (not just the forest) and specifically include the villagers."

The loping of oak and other fodder trees is usually done by the farmers to avoid injury to the trees. Agarwal (1991) found in Northern India that if careful lopping for tree fodder is practiced, overall fodder productivity is enhanced. Carpenter and Zomer (1996) warned that when trees are continuously severely lopped, dense understory such as *Mussaenda frondosa* are found which are inpalatable to livestock. Twigs and fallen branches are gathered as firewood, mainly by women. There are a few exceptions when there is illegal tree felling as I previously explained.

Common Property Resources: A Historical Overview of the *Kipat* System

The *kipat* system, the traditional form of land tenure practiced by the Rais, is an important component of their identity and historic claim as the original settlers in this area. For a detailed description of the *kipat* system in Eastern Nepal (approved during the Shah regime) in the late 1700's, refer to Chapter 1. Several interviews conducted with key informants looked at historical changes in rights and access to common property resources within a political and economic framework. An 82-year-old Yangden Rai described the early *kipat* system.

Many years ago Sherpas lived in Yangden before the Rais. Before my paternal great-grandfather came to Yangden, seven generations ago[39] *(210 years), the land in Yangden belonged to Sherpas. The first jimmāwāl (village headman) was a Sherpa named Karma Kalchong Sherpa. The jimmāwāl borrowed money from my great-grandfather but he could not repay the money on time. The Sherpa jimmāwāl told him that he could be the jimmāwāl and take the land. At that time, no one sold their land. If a person left the village, the land went to the jimmāwāl. When the Sherpas moved away from Yangden, they said to*

---

39 In these scenarios, one generation is considered to be thirty years.

the Rais, "You cannot sell the land to anybody without first asking the Sherpas." A black rooster was killed and one to two drops of its blood was drunk by both parties as an unsigned verbal agreement. If the promise was broken, the Rais felt the Bhagavan (God) would harm them. Even now the Rais still believe they can not sell the land, however, according to the present law they can. The treaty goes back many years.

Before land laws changed around 1900 under the *kipat* system, land could not be sold or alienated outside the community. This land formally could only be mortgaged according to the land reform acts. After these changes occurred, several Rais sold land to Sherpas in order to cover funeral expenses. Regmi (1978:537) points out that outside of Far Kirat, the geographical location has been assumed to be a more important factor in determining *kipat* ownership than kinship. This may be the reason why the first Sherpa in this area was a *jimmāwāl* (for an explanation of why this area is considered part of Far Kirat, refer to Chapter 1).

The historical and political system during the early *kipat* era is further explained by the villagers.

Many generations ago, farmers cleared vacant land to stake claims. In early times, taxes were paid by the clans but later by individual farmers. The farmer paid an annual tax payment of two rupees, called kharkeli, for unirrigated fields (bāri), to the jimmāwāl located in Chhoyeng, a neighboring Rai village. This tax payment ensured them the right to retain the land registered in their name and they received a note from the jimmāwāl regarding the plot of land. One sheep was also sacrificed during this transaction. After collecting all the taxes, the jimmāwāl presented it to the adda (government office for registering land). Previously Chainpur contained the land registrations, however, they are now kept in Khandbari. The Rais said they paid a livestock tax to the jimmāwāl and the rate varied according to the farmer's wealth, five paisā[40] for rich people and three paisā for poor people.

Exploitation occurred when the *jimmāwāl* was in charge because he was the most powerful person in the village and made the law (see Forbes 1993). The *jimmāwāl* was said to have cheated the villagers, because he collected the Rais' money which should have gone to the government. A 35-year-old Yangden Rai talked about the former *kipat* system during the time when a corrupt *jimmāwāl* in the area took advantage of the people.

Seventy years ago (1922), my grandfather's land was registered by the thari,[41] who worked for the jimmāwāl. During Dasaī (the Hindu festival in honor of goddess Durga Bhawani) there was a tradition where each house would give twelve pāthi (48 kg) of crops to the jimmāwāl. One member from every household contributed a day of labor to clean around the jimmāwāl's house, for example, apply cowdung mud to the ground floor. Even the poor

---

40 One hundred *paisā* is equivalent to one rupee.
41 A non-official tax collection functionary in the hill districts.

people had to pay the amount of ten pāthi of grain. At that time, a huge earthquake occurred and my grandfather died. My father was only fourteen years old and could not work in the fields. He was not able to make the ten pāthi grain payment to the jimmāwāl. The jimmāwāl came to his house and asked for the payment. My father said, "I can not pay you. Please excuse me for one or two years." The jimmāwāl became angry and hit him. My father was very "duhkha" (miserable). He then went to other people's houses to work and asked for his salary as food in kind (two pāthi of grain). Finally he was able to give the ten pāthi payment to the jimmāwāl.

The crops given to the *jimmāwāl* were referred to as *kajkalyanko walak,* a levy collected in the form of garden or other produce during festivals and other occasions (Regmi 1978:860). Also, around ninety years ago each house gave five days of labor to the *jimmāwāl.* These *rakam* obligations constituted an onerous burden on the farmers, which was only partially compensated for by minor tax concessions and tenurial privileges. Regmi (1978:529) claimed that under this system "even on regular payments of taxes in cash or kind," there was no security on the possession and use of land for agricultural purposes. The *jimmāwāl* was able to manipulate the land tenure rights.

These historical scenarios document that there was no mercy for the poor who were not given any respite on the tax payments and if they did not make them for any reason, they could lose their land. The land at the *lekh* is communally owned and neither money nor taxes are given for grazing animals at the pasture lands (*kharka*). Previously (after 1929) *ghum* (woven rain covers of bamboo), butter and potatoes were often used for payment in lieu of cash as taxes on pasture land in the hills (Regmi 1978:42). As discussed in Chapter 1, the political regime continually aggravated the hill farmers' situation by causing an increase in poverty and indebtedness.

Indigenous Forest User Groups and *Jangal* Management

Fisher (1989:1) defines a forest management system as a "set of technical and social arrangements involved in the management of forests, including the protection, harvesting and distribution of forest products." Common property management makes use of indigenous knowledge, informal user groups and religious sanctions. Jodha (1994:150-51) asserts that in India common property resources are often unnoticed by rural researchers and development planners. One reason was an "inadequate understanding of the survival mechanisms used by the poor," in addition to lack of understanding of how common property resources intersect with private property-based activities in rural areas. The author argues that the privatization of common property resources as a strategy to help the rural poor in India yielded a negative result. Common property resource activities are viewed as low pay-off options.

Informal forest committees are often formed to protect local forests in reaction to a shortage of common property resource management systems (Fox 1995). Whenever there is a problem regarding management of the communal *jaṅgal* and its resources, a meeting is called (Saisima Sherpas do not attend the village council because of its more remote location and the strict Buddhist rules they follow). Any dispute regarding the communal forest concerns all villagers. Forbes (1993) suggested *kipat* tenure is viewed as a system of local self government, although the "legal" owner of the land is now the national government. Mainly men were in attendance at these meetings. Women usually attend the meetings only if they are the head of the household or if their husbands are absent from the village. Usually the eldest member of the cluster of villages is in charge of the meeting.

The use of the *jaṅgal* is controlled by closing an area to grazing at certain times of the year to permit the protection and regeneration of specific resources. The rotational system allows the resource to grow again before it is harvested. Arnold and Campbell (1986: 447) attributed that "this system is the most important silvicultural treatment required by most community forests, including those under explicit traditional management." A shortage of *māliṅgo* (*Arundinaria aristatla*) is noted as one of the biggest problems in the area. The following anecdote describes how the conflict between livestock and the communal *jaṅgal* management of *māliṅgo* was resolved.

> *Four years ago (1988), māliṅgo became scarce. A meeting was called for all villagers in Yangden, Dobatak and Gongtala who use these resources to attend. Because young māliṅgo is also a source of cow fodder, there was a general consensus that all cows must graze in another part of the jaṅgal until the māliṅgo became mature enough and was no longer edible. Also, the Rais who are very dependent on māliṅgo for making woven mats (bhakāri) were asked not to collect in this area. A forest watchman was chosen to impose fines on whoever would break the agreement.*

The villagers cooperatively proposed a solution to the resource depletion and enforced it by the willing consent of herders to keep their animals out of the area and the fear of a fine as punishment. This approach is less restrictive than the fencing of livestock suggested by Rajbhandari (1991) who views free grazing as a destructive aspect of grassland management.

Stephen Bezruchka

Mālingo (Arundinaria aristatla) collected from the *jangal*.

Gilmour (1989) asserts that local villagers will adopt rules or measures to manage a resource, mainly the larger trees, only when it becomes scarce or valuable. In this area, it has not been the case for all farmers because the daily rotation of herds is based on the decision-making consensus of Rai and Sherpa herders. The group's decision on where to graze the livestock depends on fodder availability. In all the villages, a yearly meeting is held prior to the *lekh's* summer transhumance. The villagers collectively decide where to graze the high-altitude livestock and how the pasture land will be allocated in order to ensure fresh pasturage. Controlled grazing helps to ensure protection of the resources.

These traditional tenure systems are important to prevent a shortage of forest products and conserve forest resources by limiting access and imposing restrictions. Trees are not cut on community land and only branches and dried wood can be collected. Before felling a tree to either build a house or to stockpile a wood, permission is required from the village chairman (*adhyaka*). A meeting is called, and afterwards, all the household heads must give approval. If a farmer does not follow this practice and fells a tree, he is fined Rs 500. Social sanctions and fines help to prevent an infringement of village policies and maintain social harmony.

The economic value of a resource may act as an incentive for the villager's adoption of rules for communal use, as is the case for *chiraito*, discussed in Chapter 8. *Chiraito* grows naturally on swidden cultivated fields and if the plant is picked after it has matured, the seeds drop and regeneration occurs. A Sherpa farmer said, "if the market price for *chiraito* becomes high enough, we may consider planting it in large quantities." The "may" leaves open the question whether forests will be exploited for monetary gains (Gilmour 1989), especially if the farmers feel threatened by external agencies which may undermine tenure security, or whether they will develop rules to manage it.

Land Resource Rituals

Local knowledge systems include mechanisms of ecological adaptation bound up in rituals. Both the Sherpas and the Rai religious beliefs and sanctions have important implications for their interaction with the environment and in promoting conservation.

Indigenous religion can have a positive impact on biological diversity (Ingels 1995). The cognized maps of the Rais and Sherpas are expressed not only in their farming systems but also in traditional rituals which were discussed in Chapter 6. Buddhism stresses the theory of non-duality, that is, humans are a part of nature and therefore we should treat "nature" as we do ourselves. Rules governing behavior help to reinforce the notion that human activity should be carried out with a feeling of conservation, not of exploitation (Eppsteiner 1988). Stevens (1993:325-6) supports this when he argued that "one lesson has been that neither the permit system of tree felling in use from 1965 through 1976 [Forest Nationalization Act] nor the National Park ban [Sagarmatha National Park] on tree felling has been as fully effective as Sherpa religious beliefs, self-restraint and community vigilance in protecting sacred forests." Aris (1990) illustrates that local awareness of tree preservation can be traced back at least to the 11th Century by quoting Milarepa (Tibet's beloved poet-saint).

> *In the south lies Nepal, the land of rocks and thunder.*
> *If the natives with their axes*
> *Cut not the healing tree of sandalwood,*
> *The Mon intruder will not harm them.*
> *To preserve the woods among compatriots*
> *Is of great importance.*
> (Milarepa 1040-1123)

Both sacred forests and sacred groves are referred to by different authors, and in my discussion among Rai and Sherpa villagers, I use Mansberger's (1991) definition. A sacred forest "is usually an extensive area preserving some or all of its primitive wildness, often with wild animals in it," while "a grove is usually not a very large area, comprised of clusters of trees often cleared or under brush and can be wild, tended or cultivated" (Mansberger 1991:44-45). Carpenter and Zomer (1996) found that *raniban*, "queen's forest" (small patches of late-successional subtropical forest) in the Makalu-Barun National Park and Conservation Area were protected by religious custom and local laws. The authors asserted that important ecological functions, for example, slope stabilization or watershed protection are attributed to sacred forests. Sacred groves are also a site for religious rituals, conflict resolution and a reservoir of biological diversity (Schelhas and Greenberg 1996). In the sacred groves, all forms of vegetation are protected by a deity.

According to Gadgil and Vartak (1994: 82), the sacred groves "harbor vegetation in its climax formation because they may represent the only type of forest in near virgin condition." The authors associate sacred groves with hunting and gathering groups. In some parts of India, the sanctions on sacred groves have been weakened and tree felling has occurred, occasionally resulting in a total degradation of the sacred grove (Gadgil and Vartak 1994). Stevens (1997) reports that in Khumbu sacred forests, tree felling and tree lopping is strictly prohibited, however, grazing and the collection of dead branches off the forest floor is permitted. Rules regarding resource use in sacred forests thus vary according to the social group consensus, ecological and economic constraints, etc. Mansberger (1991:245) discusses the concerns among respondents in Kathmandu Valley of the consequences of green shopping. *Tejpatra* (*Cinnamomom tamala*) was found in sacred forests in the past in "great numbers," but once "people started to realize its market value, the tree got wiped out."

This is not the case for the sacred forests in this area where religious sanctions are still strong and strictly adhered to by both Rai and Sherpas (This does not hold true, however, for Rai hunters from neighboring villages). Surrounding the Rai and Sherpa villages, there are "sacred" forests, referred to as *devī than* (place of the goddess). A person can only enter a *devī than* when making a *pūjā* (ritual sacrifice) for the goddess. Hunting, grazing livestock, felling trees and the collection of firewood, medicinal plants and bamboo is prohibited. As one Rai said, "even if there is gold in the *devī than*, nobody can take it." The farmers said if trees are cut, the goddess will get angry and cause hail, rain and high winds to come and result in landslides. Similar findings were reported by Mansberger (1991:144) who described the *ban devī's* role as essentially that of a "divine forest ranger(ess)."

Women herders and children, ten years and older, form groups and decide daily where to take their livestock for grazing, depending on fodder availability. Also, when fodder must be collected for stall-feeding, it is done by groups of women. The *jańgal* is "lonely" (Carpenter 1991) and envelops people in a sense of fear because "it is a habitat for *bāgh*[42] and other wild animals." I was shown fresh "*bāgh*" tracks on the footpath in the *jańgal* by anxious Sherpa female friends when we were collecting fodder.

Sherpa women carrying baskets of fodder

The Rai and Sherpa farmers drew natural resource maps of their villages and the surrounding *jańgal* to show where they collect firewood, *sisnu*, *chiraito* and the location of the sacred forests (see Figures 8-11). Nazarea-Sandoval (1995:103) notes that a mental map is a "graphic representation of the cognitive image pertaining to a certain place or location." She remarked that people tend to distort proportion and scale according to how they view specific objects. Figure 8 shows how the cultural landscape in Gongtala has changed because of trekking opportunities (modern houses outnumber traditional houses) and the increased cash flow to the village.

The "hotel" in Figure 9 was drawn much larger than its actual proportion to the other houses in the area. However, this may be because it represents the onset of tourism to the area and the ensuing income for the enterprising Sherpa who had it built. On the other hand, even though the Sherpas view *sisnu* as low status and do not serve it when there are guests, it was included on all the participatory maps showing its importance for the households. *Chiraito* sales were the largest in the Dobatak area

---

42 *Bāgh* means tiger in Nepali. However, the villagers use *bāgh* in a general way to refer to any large wild cat, including leopard which occurs commonly throughout Nepal.

(see Chapter 8). Yak horns are a symbolic religious marker often found at the entrance of Buddhist monasteries and sacred caves.

Figure 8. Natural resources identified by the Sherpas in Gongtala

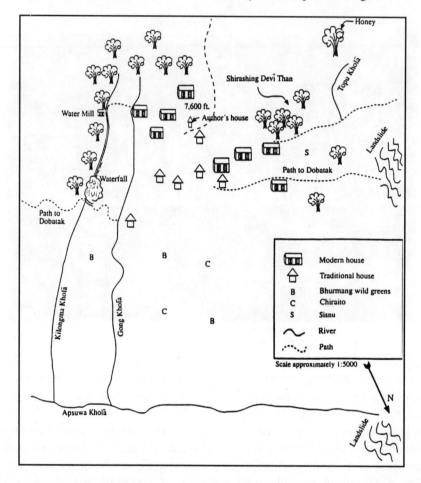

In Figure 10, *chiraito* is not shown because in the area of Saisima swidden cultivation is not practiced. *Lokta* is the main *jaṅgal* resource sold for cash. 'Wild' honey is found in the precipices above the village and only a few Rais who are *dhāmīs* are known to be able to extract it.

Figure 11 shows how widely the Yangden houses are scattered. It took one hour to walk from Jisdo *ṭol* to Yangden *ṭol*. In drawing maps people are said to draw features of the environment that have a functional relationship to behavior. For example, the *ban jhākri*, discussed in Chapter 6, was drawn near the rivers, as it is one of the places where it resides. The *ban jhākri* was not included in the Sherpa's participatory maps. As B. Gurung (1995) points out, because the local people's spiritual beliefs, cosmologies and world views are an integral part of the natural resource management system, outside actors need to understand their importance. Of all the participants, the Rais named the most resources, which shows how they are dependent on them.

Figure 9. Participatory map of the natural resources in Dobatak

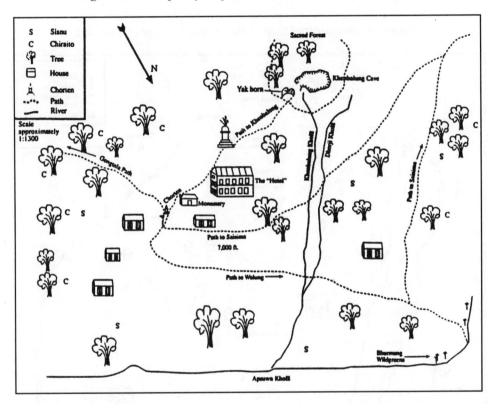

The Sherpas living in Saisima said that wildlife is abundant in the sacred forest. The head *lama* has prohibited the killing of wild animals in the surrounding valleys of Saisima. The passages of Buddhist law within the Tibetan text concerning Khenbalung show how the unique nature of this sacred area has been sustained.

> *You should make laws for Rig-gya (prohibiting hunting) with the (upper portion of the) valley. This valley is the place of Chenrezig (Avalokitesvara), so you should increase white (good) dharma. You should not annoy the spirits who live in the jungle and rocks*
>
> (translated in Reinhard 1978:21).

There are exceptions to these rules, however, and as one Sherpa said, "we sometimes allow Rai hunters into Saisima because the wild animals eat our crops and we want them killed." The Sherpa's remark does not invoke the image of the ideal

Buddhist. As Ramble and Chapagain (1990) explained, religious strictures bend because of economic necessity (see Chapter 2). Shortly before the onset of *Dasaī*, several cows "fell" and died in the *jaṅgal*, in Sherpa areas where there were no cliffs. The 82-year-old Yangden man said, there are no guns in Yangden and our people do not hunt. The hunters come from other neighboring villages and sometimes as far away as Solu Khumbu.

Figure 10. Participatory map of the natural resources in Saisima

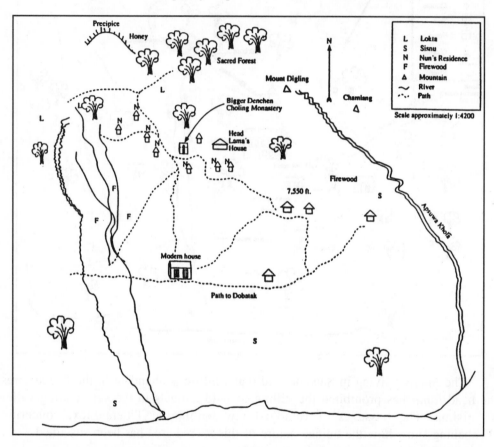

I observed a form of natural resource conservation when I accompanied the Rais to their summer sheep shelter at the *lekh*. *Pūjās* are held daily by a priest (*pūjāri*) once a particular location is reached where the Rais believe numerous deities reside in the mountains, forests, rivers and sky. These *pūjās* are conducted for the deities in order to ensure that the Rai's sheep as well as their own welfare will be protected. The *pūjā* is a propitiatory ritual for the deities and also expresses conservation ethics. The shepherds fear that the deity will become angry and harm them if they abuse the environment.

Figure 11. Participatory map of the natural resources in Yangden

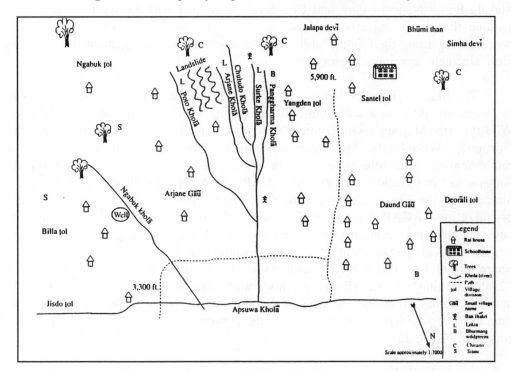

A Rai man had been cutting trees for a Sherpa man near Gongtala. He called for a *dhāmī* to determine why he had severe pain in his right arm. The *dhāmī* conducted a rice divination and said, "you have been cutting the trees in this *jaṅgal* too close to the village and have angered the *ban jhākri*." This incident shows how the *ban jhākri* acts as a symbol to inhibit deviant behavior and indirectly preserve the natural resources.

External Actors

National Parks

Dove and Rao (1990) assert that protected areas need local participation in order to incorporate indigenous knowledge systems, traditional (social and religious) institutions, as well as address the basic needs of the local people. The recent trend is to instill "new partnerships" to ensure that the project will benefit the inhabitants and become self-sustaining (McNeely 1995). Nepal "has become a leading innovator" in the establishment of protected areas co-managed by indigenous communities (Stevens 1997:63).

I lived in the area from 1991 to 1993, prior to the implementation of the Makalu-Barun National Park and Conservation Area. I will not try to evaluate nor critique the park's objectives. However, I will try to express the local people's concerns regarding their future ability to the use and the management of what they view are their "communal resources."

The Makalu-Barun National Park and Conservation Area, formally established in December 1992, is a collaborative undertaking by the Department of Parks and Wildlife, His Majesty's Government of Nepal, and The Mountain Institute, USA (formerly Woodlands Mountain Institute). The Makalu-Barun project is implementing a people-national park management approach modeled on the Annapurna Conservation Area Project (ACAP). ACAP was established in 1986, and is managed by the King Mahendra Trust for Nature Conservation, a non-governmental organization. ACAP has a buffer zone of conservation around a core of environmental preservation (natural reserve) which relies on a grassroots approach based on local management of natural resources and participation (Stevens and Sherpa 1993:74). The MBCP encompasses 1,500 km2 of national park and has 32,000 inhabitants living within the conservation area or buffer zone of 830 km2. Its main objective is to protect the biological and cultural diversity of the area at the same time encouraging active participation by the communities residing within the buffer zone (see Figure 12).

Figure 12. Makalu-Barun National Park and Conservation Area and other national parks and reserves in Nepal (Stevens 1997).

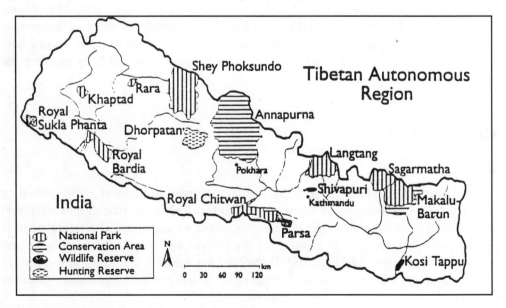

Rai and Sherpa Perceptions of the National Park

The Rais and Sherpas have different perceptions of the National Park based on different categories of basic needs and use of the *jaṅgal* resources. I had not planned to study intra-or inter-ethnic inequality, that is social asymmetry, but it arose out of the context of *jaṅgal* use.

The Rais are wary of the national park and the Cadastral Survey, perceived by the local people as exerting "outside" control and restrictions on their *jaṅgal* resources. Ramble and Chapagain (1990:24) remarked about similar responses when the plan of the *rastriya nikunj* (national park) was discussed with the local people. The researchers said that the word *rastriya* created suspicion "towards what they suspected would be an infringement of hereditary rights by the government." It should be remembered that the Rais "claim to be the original inhabitants" of Majh-Kirat, often referring to other groups as immigrants (Sharma et al 1991). *Kipat* rights have given them power and control of the *jaṅgal* resources which have been controlled and managed by indigenous and traditional forest-user groups. Their concerns must be understood within the overall historical context of the previous monarch rule which impoverished the peasants (see Chapter 1). The Rais expressed their concerns about the newly established Makalu-Barun National Park.

> *This land belongs to us and is registered with the government. We have paid taxes on the pasture lands. We worry that the bears and tahr will graze freely and eat our crops. Under the National Park system we feel we will not be able to cut māliṅgo. We need sisnu to feed the pigs and as food for our families. We need firewood and to be able to continue to graze our animals in the jaṅgal and the lekh. We must continue to do khoriya (swidden cultivation) because if we don't, we will not have enough food. Also if khoriya is not done, the jaṅgal will grow big and there will be no chiraito. Income from chiraito helps us to buy food for the periods when our food stocks are depleted.*

A contrasting and more positive view of the national park was expressed by several of the wealthier Sherpas. They feel the park will provide benefits from an increase in tourists to the area and forsee opportunities to work as porters and guides in the tourist industry. The Sherpas were traditionally pastoralists and long-distance traders until the border with Tibet closed in the early 1960's (Furer-Haimendorf 1975). They soon adapted to the new economic opportunities brought about by tourism beginning in the Mount Everest region when Nepal opened up to foreigners in 1950. Sherpas who work as high-altitude mountain climbers and trekking guides have reaped benefits from their association with foreigners (Brower 1991). Sharma et al (1991:14) have asserted that Sherpas pick up new ideas and innovations more quickly than most other ethnic groups. In my area, the Sherpas are already experiencing socio-economic changes, and many new houses have been built of stone and wood. The cultural landscape of Gongtala, Dobatak and Saisima is beginning to resemble that of the Khumbu area frequented by tourists for three decades.

Brower (1991) has pointed out that along with the benefits incurred from tourism there are also costs. These include: inequity in the distribution of benefits from tourism, shortages in the labor pool for agriculture and animal husbandry (and women bearing the brunt of the increased work load), and inflation for necessities, including foodstuffs, which affects local people and tourists equally. For example, when I trekked westwards from Gongtala to Junbesi in the Solu region (south of Khumbu), I priced an egg. In Gongtala, the price of one egg was two rupees and in Junbesi, a tourist destination on the route to Khumbu, it was twelve rupees. Lastly, there can occur a disruption in the traditional systems, for example, outward migration, a decrease in agricultural production and a loss of traditional knowledge.

One Sherpa living in Dobatak was building a "hotel" because he was anticipating the influx of tourists to visit Khenbalung (the cave where Padmasambhava had meditated) and the ensuing income this would elicit. At the same time, he was apprehensive about the national park because he treks throughout Nepal and was aware of the people-national park conflicts occurring elsewhere. He spoke about the army stationed in Sagarmatha National Park (Mount Everest) to enforce regulations of firewood collection and restriction in grazing areas and feared that it would happen in this area. In Sagarmatha National Park (established in 1976), numerous conflicts have occurred between local people and park authorities in the beginning phases of its implementation through the 1980's. These conflicts arose over the issue of forest use and management. The stationing of non-Sherpa army personnel in the park as guards had angered many people. The traditional elder councils who had traditionally managed the national resources dissolved in the 1950's (Stevens 1993; Weber 1991). Only when village forest management committees were established and the Sherpas felt the forests were to some extent once again their own did tensions ease up (Stevens and Sherpa 1995). These people-park conflicts are not unique to Sagarmatha but have occurred and are occurring worldwide (see McNeely 1995). Addressing the differences in interest between the local communities and conservationists in solving these conflicts is of the utmost importance.

An increase in tourism creates a pressure on the environment and sets the stage for culture change. Tirtha Shrestha stated during an interview, that tourism is seen as a potential source of funds, as well as a potential threat to the ecology of an area (Ramble 1993:38). Bunting et al (1991:162) asserts that the income generated from tourism is somewhat deceptive because only 20 cents of every three U.S. dollars spent by each trekker daily stays in the village. Tourists consume a large amount of firewood, for example, a two-month climbing expedition uses almost twice the amount of firewood a Sherpa family would use in a year. Kerosene has been made mandatory for trekking groups as part of the ecotourism approach being promoted throughout Nepal and has helped to remedy the fuelwood crisis. Yonzon and Hunter (1991) warn that tourists also create a demand for dairy products, particularly cheese, which leads to larger herds and over-grazing. The wildlife habitat in the area is encroached upon and species are threatened, such as the Red Panda in Langtang National Park, Nepal. The authors proposed as a solution to restrict the number of *caūrī* (yak-cow hybrid) and reduce cheese production. The price of cheese would be increased which would be paid for by the tourists in order that the farmer's income could be maintained.

The hotel in Dobatak

Social Asymmetry Among the Rais and Sherpas

It is important to look closer at why the Rai's and Sherpa's views of the National Park vary. The Rais previously had unrestricted access to *jaṅgal* resources (firewood, animal fodder, food and medicinal plants, building and household materials) because of community-owned *kipat* land and are more impoverished than the Sherpas. They are therefore more dependent on the *jaṅgal* and its resources as their "means of survival." The "key" to the resistance directed at the National Park is *chiraito* which grows naturally in fallow swiddened fields. The National Park will "regulate slash-and-burn practices" (Shrestha et al 1990). Swidden cultivation is extensively practiced by the Rais because of the diminished size of permanent field farm holdings, and high returns to labor, especially because many villagers are migrating for wage labor jobs. For these reasons, they are more dependent on it for subsistence than the Sherpas. A restriction in swidden cultivation practices will reduce *chiraito* and lead to a loss of an important source of cash income for the household.

The extent that these cultural perspectives are not shared by Rais and Sherpas is according to Nazarea-Sandoval (1995:xi), based on the standing of individuals "in the hierarchy of economic and social relations." The author adds that two important dimensions within the social matrix are gender and class. Thomas-Slayter and Bhatt (1994) caution that the careful planning of village-based development projects needs to pay attention to the embedded social, and particularly gender-based relationships, and to consider how they may be affected by economic and ecological transformations. Milton (1996) dispels the image that cultures are bounded entities. In societies experiencing modernity, ethnic identity can be viewed as changing along a continuum. The Sherpas can be viewed as about half way along this continuum, changing their values and beliefs (as evidenced by Sanskritization) and becoming more involved in the global process (tourism, trekking) to take advantage of economic opportunities. The Sherpas have more opportunities for a trans-cultural discourse, described by Milton (1996:170) as "a mechanism through which specific cultural phenomena are communicated beyond their location of origin." The Rais are at one end adhering to a strong commitment to the traditions of their group, and tend to be more tied to their territory.

Peluso (1992) posited important questions that need to be addressed in this situation: As for equity, how is it to be measured when different groups of users exploit different resources in conflicting ways? What are the boundaries of the set of potential users and available resources? Forbes (1993:2) conducted research in Hedangna, located in the national park a three-day walk east from my research area, and discussed how "*kipat* tenure is a map of social relations and land claims, a map that is most significant for what is left unmarked, for the absence of lines and the fluidity of boundaries." The *kipat* tenure map has a different perspective for the local population than that of the maps drawn by external agencies.

Nepali (1993) pointed out that 52 percent (122) of the local development projects in Eastern Nepal failed because of the following cultural reasons: close kinship ties, land ownership claims, failure to identify locally-felt needs, ignorance of cultural values. In these instances, outsider's perspectives outweighed those of the local people. A major finding of the Kosi Hill Area Rural Development Programme (which includes the Sankhuwasabha District) was that a household's access to extension services and credit is determined largely by the wealth and social status of household members, even when analyzing the data by distance from service point. The more well-off households have been the first to receive the greatest benefits from new development opportunities (Nabarro et al 1987). Similar findings are reported by Thomas-Slayter and Bhatt (1994). Cernea (1992) discussed how rural development projects in Pakistan had been plagued by large landowners over-exploiting government resources.

Government Policies for Natural Resource Management

Peluso (1992) has asserted that political economic structures strongly influence patterns of forest or resource access, control and management. These regulations affect the types of forest land classification used, that is, the establishment of forest protected areas, and regulations on land tenure arrangements; and hunting restrictions. These governmental regulations create permanent boundaries. A shift in tenure away from common property ownership can lead to a loss of *jaṅgal* resources and the benefits they provide for the community. Alcorn (1996:233-234) says "security of tenure is critically important for conservation, because it provides incentives for people to forego short-term gains in order to maintain forest or other habitats for future use."

The Cadastral Survey and Regulation of Swidden Cultivation

At the time of my research, the Cadastral Survey was being carried out for the first time by the Nepalese government. The survey was designed to measure, record, and map all land throughout the districts of Nepal, because most areas had not been legally surveyed and demarcated. The Cadastral Survey would clarify an individual's legal status to titles and boundaries, develop dependable agricultural statistics and provide a basis for equitable land taxation (Regmi 1978). Once the Cadastral Survey was completed, however, the farmers felt they would have to pay higher taxes to the government. Collecting land-use data was threatening to the villagers. I observed this when I was conducting the household survey and decided not to evaluate land ownership.

The Cadastral Survey began in the Sankhuwasabha District in the spring of 1994. The consequences of swidden cultivation will be discussed because external authorities judge that current users of the land are mismanaging natural resources and it is in the national interest to restrict swidden cultivation (Ghirmire 1992). These measures show a lack of understanding of the local socio-cultural and ecological systems, underestimating the value of the natural resources for the local population (Dove 1985).

Gibbons et al (1988) postulated that prior to the Cadastral Survey, grazing and forest land would be brought under cultivation by farmers in order to establish land tenure rights, because land that is not cultivated can be claimed by the government. This is a widely held view in many countries under colonial as well as post colonial regimes (Dove 1993b), and is vividly described by Elwin (1994) for the Baiga of Central India. This was consistent with the view held by the farmers in this area in 1992 who felt that unless forest could be claimed as cultivated, the land would not be recorded as private property. The Cadastral Survey would then claim this land as property of the government. Nepali (1993:244) points out, "although it appears that it is the absence of a Cadastral Survey that prevents claims over land from being substantiated, in fact, due to government land policies even the Cadastral Survey will not determine ancient claims over the land." Once the Cadastral Survey has been completed, the *kipat* system will be totally abolished. The Cadastral Survey was an issue of conflict and exploitation for the local villagers as narrated by a Sherpa farmer who lost a great deal of land because of it.

> *Recently the Cadastral Survey came to measure my land at Gupha Ḍāḍā. My father bought this fertile land from a Tamang eighteen years ago for thirteen hundred rupees.. I lost my land because there were not enough family members to cut the trees on my land so after about eight years, the jaṅgal grew again. The land surveyors said I had more land than I needed and took it away even though I had a paper showing it was my land. They said if I paid them ten thousand rupees, I could keep the land. The money would not be given to the government but would go into the surveyor's pockets. I don't feel anything as I have to follow the government rule and was too late to cut the trees.*

This anecdote illustrates why the Cadastral Survey was a direct cause of increased deforestation throughout Nepal. For swidden cultivated fields to regain fertility, an adequate fallow period is required. On the contrary, if the people feel threatened that they may lose the land to outside authorities, if it appears as "*jaṅgal*," the land is swiddened early. The soil is not fertile, the crops will not grow well and the result is less food brought to the family hearth. Another reason the fields are left fallow is that there are fewer adult laborers due to seasonal and long-term migration, and family members are not able to compete with the weeds.

Slash-and-burn prior to the Cadastral Survey

Many burning patches of slash-and-burn land were visible during a brief visit I made to the area in the spring of 1994. This scenario seems to repeat what occurred in Nepal when the government nationalized forests in 1957, to curb deforestation and converted communal forest into *de jure* state property. As Feeney et al (1990:8) concluded, the aftermath of these laws "approximated the creation of *de facto* open access" by villagers who lost control of nearby forests.

Chambers (1979:3) describes the asymmetry in power in the following quote, "We are confronted again with who is powerful and who is weak. Rural people generally are weak vis-a-vis officials and professionals; and rural society is itself differentiated into those who are stronger and whose interests and capabilities are liable to dominate those with less voice and control."

Divergence of Cultural Perspectives and the *Jaṅgal* Landscape

Landscapes and environments "are perceived and interpreted from many different and contested points of view which reflect the particular experience, culture and values of the viewer" (Blaikie 1995:203). Government officials, development planners and scientists have a nature conservation perspective to protect biodiversity and separate nature from processes and products of human activities (Milton 1996). "Nature is at its most valuable when it is untouched by human hand" (Milton 1996:124). This paradigm was held by conservationists in the early stages of developing national parks, and a classic example is Yellowstone National Park (Stevens 1997). On the other hand, the farmers co-exist with nature using their intricate indigenous knowledge systems.

Seeland (1997:15) stressed that policies have to account for "different cultural views of nature and environment and their respective economic needs and political interests." The sociology of knowledge held by outside versus local actors is a complex, evolving issue. A vivid example is described by Gilmour et al (1989) who discussed contrasting views of "forest ownership," by villagers and forest department staff which affected actual decision-making about resources used by them.

Different versions of environmental issues emerge depending on who the actor is and the type of value of the resource, that is, use versus subsistence. A plurality of points of view exists about the environment and different groups may covet the same resource for different reasons, which can lead to conflicts (Seeland 1997). For example, outside actors or elites within the community may notice the valuable medicinal plants for export purposes while the villagers, particularly the poor, may see the medicinal plant properties for treating household member illnesses and "weeds" as valuable.

King et al (1990:10), referring to the formation of community forestry-user groups, discussed problems of communication between different management levels which can lead to misunderstandings. As the authors saw it, participants "talk past" each other or the flow of information may either be in one direction, or unbalanced. The authors suggested re-training the lower ranking forest staff skills so they can effectively communicate with the villagers.

Men often migrate, creating a shortage of farm labor, and women must assume an increased burden of responsibilities (and time), including increased decision-making for the household. Because the Nepali language, the *lingua franca*, is spoken by a minority of the women, it can present a disadvantage for them when conducting trade and credit relations. These women are the key gatekeepers in the foodpath and have the primary responsibility for gathering *jaṅgal* ingestibles, collecting fodder, small firewood (twigs), litter for animal bedding, and childcare (medicinal plants). They are wealth holders of indigenous knowledge regarding the *jaṅgal* resources. This knowledge is context specific even for the youngest social group (see Daniggelis 1994a; Saul 1994). As mentioned earlier, village meetings are usually dominated by men, although women use and manage the resources vital for their survival. Therefore they know the current situation of the resources and need to become active participants at the meetings so their voices will be heard. The constraints to women's involvement in community forestry groups and their successful participation on forestry committees is discussed by Denholm (1991), D. Gurung (1987), Pandey (1987), Prasai et al (1987), Seeley (1989).

The villagers are motivated to improve the future for their children. In the first phase of the MBCP village incentive program, the community in the project area was given Rs 6,000 to improve the school in Yangden. Every household in Yangden, Gongtala, Dobatak and Saisima contributed an additional Rs 150 to match the amount that was given by the Makalu-Barun Conservation Project. For a poor villager, this is a large amount of money to provide when the main means of exchange is still bartering. Each household also contributed one day of human labor for rebuilding the school.

Solutions are recognized as having **both** environmental and social dimensions because what people tend to pay attention to is "culturally defined." The sociology of knowledge is worthy of attention since it is tied in with reality which often reflects strong social and political forces (Dove 1993). Within their economic constraints arising from poverty, the local people live in harmony with nature. For generations the Rais have been taking care of the land through common property management and conservation of the forest and pasture lands (Nepali 1993). For the Rais and poorer Sherpas, the land is vital in coping with their precarious food situation. They use an extensive and complex ideational system of social values, culture and religion to sustain the natural resources and biodiversity. When confronting the issue of poverty and problem of food security, it is vital that programs and policies address women's role in meeting basic household needs. The distribution of benefits will then be targeted to include all socio-economic, ethnic and gender groups. These issues are addressed by Nazarea-Sandoval (1995).

> We cannot deny the boundedness of the system in which day-to-day agricultural decision making takes place or the fact that the ability to recognize the existence of alternatives and exercise choice is directly proportional to the individual's standing in the hierarchy of social and economic relations. This is so not only because access to resources is systematically skewed but even more because the distribution of knowledge is socially patterned and culturally rationalized (Nazarea-Sandoval 1995:141).

CHAPTER 10

Is the *Jaṅgal* Really Hidden?
And From Whom?

This book has viewed how marginalized impoverished communities living on the fringes of the *jaṅgal* interact with it. I found the *jaṅgal* to encompass not only trees, but also fallow land, brush land, along with the high pastures (*lekh*), the resources that **lie in and under the trees and are oftentimes "hidden" from outsiders**. While some people have looked at nutrition, or medicine or fodder, my work shows the *jaṅgal* (and *lekh*) to be useful in a much broader way than is any part of the agricultural land. Part of the problem is in defining the agro-ecosystem as solely the farm fields with domestic crops. This is readily observable, but is only a component of a much larger food production system. The non-farm environment is the ultimate polycrop system. Far from being the residual category implied by using terms such as non-timber resources, the *jaṅgal* is the ultimate resource base – as we should have known, since it has the maximum diversity.

The most stable resource system that the foraging farmers have owned is the *jaṅgal*, consisting not only of trees for building timber and bedding fodder for livestock, but also serving as a repository of many valuable nutritional, medicinal, economic (subsistence and cash), religious and cultural resources. Because these resources may be found in areas consisting only of shrubs, they are often disregarded as 'weeds' by other social and political groups. Cultural models of both the Rai and poorer Sherpa farmers who eke out their livelihood by relying on the *jaṅgal's* sustenance looked at the full value of it.

Previous research on the forest has tended to look at narrow aspects of foraging, even though it encompasses a vast domain. *Jaṅgal* foraging enters every aspect of the farmer's life, and the poorer the farmer is, the more this holds true. Now there is a resurgence of dependence on the *jaṅgal* by farmers because of an increased fragmentation of household land, increasing indebtedness and poverty. One hundred years ago this was less important. However, at the present time the poor farmers are forced into an increased dependence on the *jaṅgal* at the same time that it is disappearing.

Throughout the history of Nepal, the Rai farmers, along with other Tibeto-Burman groups, have been pushed to more marginal and infertile lands by the Indo-Aryan ruling powers. Foraging and farming are complementary systems. The Rais and poorer Sherpas use a broad spectrum of coping strategies to buffer the food scarcity period. These means consist of the extraction of *jaṅgal* ingestibles for consumption, including preservation and storage for *anikāl* (annual food scarcity period); the trade in medicinal plants, particularly, *chiraito* (*Swertia chirata*), and shifting cultivation. In addition to these strategies, the farmers also employ social networks, wage labor or food-in-kind payments, short or long-term migration and loans, to buffer *anikāl*.

The impoverished hill farmers live a day-to-day existence in a very unpredictable environment which can destroy their entire crop overnight. They rely on a wealth of cultural knowledge which has been transmitted inter-generationally and adapted in order to allow flexibility and quick response in adjusting to these unexpected crises. They believe that their own destiny is controlled by deities or spirits which reside in the *jaṅgal* and therefore take great care to appease them. Heavy rains, strong winds and landslides are believed to be caused by the wrath of an angry deity or spirit which can wash away a household and land in the middle of the night.

The Rai View of the *Jaṅgal*

The term *jaṅgal* encompasses much more to the Rai farmer than the earlier views described in Chapter 1. A quick review of them will help to emphasize my point. For the wealthy high-caste Hindus, the *jaṅgal* is "polluted," "uncivilized" and "wild, and therefore needs to be controlled and tamed. The Levi-Straussian structural approach separates culture from nature, with the former having the power to dominate and transform the latter. The Sherpas contrarily see the forest as uncontrollable and sacred. None of these views represents what the *jaṅgal* means to the poor foraging Rai farmer who, through the centuries, has maintained a mutualistic relationship with it. The Sherpa view has the most affinity with that held by the Rais, in that the Rais consider the *jaṅgal* as sacred and uncontrollable. The Rais differ from the Sherpas because they maintain a strong ethnic identity with the *jaṅgal*. For these reasons, the Rais have carefully used and managed the *jaṅgal* and its resources in a carefully controlled way, because it represents their key to survival. This I will explain further on in this chapter.

The Rais, claiming to be the original inhabitants of the area, have lived and survived on hunting and gathering and swidden cultivation for centuries. The Rais's historic ties to the *jaṅgal* are well known because from earliest times they have been known as great hunters, warriors and traders of forest products with lowlanders. The terms, *śikāri* (hunter) and *jaṅgali* people, are still used by high caste Hindus and Sherpas to describe the Rais, often in a derogatory manner.

Oral history can contribute to a better understanding of the historic beliefs and motivations held by a group of people which can help explain persistence of a tuber, *ghar tarul* (*Dioscorea versicolor*), as a cultural identity symbol among the Rais. An elderly Rai informant revealed that one of the main reasons his ancestors (the Kulunge Rai) settled in the area now known as Yangden was that the edible tubers, *ghar tarul* and *ban tarul* (*Dioscorea alata L.*) were plentiful in the *jaṅgal*. In their former geographical settlement of Mahakulung, food was scarce and the people may have been aware that in case of a crop failure, this 'wild' food source would help pull them through the hard times. Seven generations later, the symbolic significance of *ghar tarul* is still apparent in the *Maghe Sankrānti* celebration held annually by the Rais to mark the agricultural new year. During the pre-monsoon period, when food items are in short supply, the Rais dig up *ghar tarul*, which is of high nutritional value, contributing not only energy, but valuable protein and iron as well. For example, *ghar tarul* provides twenty kilo-calories more per 100 gram portion than an equivalent amount of their common staple maize, and 50 kilo-calories more than potatoes for the same amount. *Ghar tarul*, when compared to all other 'wild' and domesticated tubers which have been analyzed, ranks the highest for protein and iron (twelve times the amount in carrots) and is second to carrots for calcium (refer to Chapter 7, Tables 33 and 34). Thus *ghar tarul* may be the major source for both protein and iron, especially because dairy products and legumes are in short supply at this time of the year, and poor Rai farmers own few milk-producing livestock. This is an excellent illustration of how scientists can learn from the local farmers who have adapted to their situation for centuries, and alert them to highly nutritional, but devalued foods.

Not to be left out of this discussion is the religious importance of the *jaṅgal* for the Rais, the home of deities and protector spirits. The *jaṅgal* is the glue that helps them maintain their ethnic identity. Gaenszle (1992:4) discusses the persistence with which the Kulunge Rai adhere to their cultural tradition and maintain their ethnic identity by comparing them with the neighboring Mewahang Rai (the second eldest brother in the origin myth) who live in more fertile areas.

> The Kulunge Rai "tend to dominate in the domain of shamanism, and in fact, they often claim to be the ones who stick more closely to their tradition and have more ritual and mythological knowledge (and there is in fact some truth in this). The Kulunge generally still speak their own language whereas many Mewahang are no longer able to speak theirs. So it seems as if the Khambuhang [mainly Kulunge] compensated their political inferiority by stressing their cultural identity." Gaenszle (1992:4)

The *ban jhākri* (see Chapter 6) symbolizes how forest degradation can have harmful repercussions, for example, in the form of landslides. Thus tree cutting in or near sacred areas is negatively sanctioned and is a local means of preserving biodiversity. Figures 6 and 11 attest to the continued presence of the *ban jhākri* and its association with the surrounding natural areas. The Rais pay homage regularly to the *ban jhākri* and residing deities, in order to maintain harmony which encourages conservation of the 'wild' environment, the home of these deities.

Clashing Cultural Perspectives

The Rais and Hindus hold contrasting cultural perspectives toward the same environmental phenomena, the *jangal*. For the Rais, it is sacred, as well as economically valued, while for the Hindus,[43] it is polluting and nearly useless. The Hindus have migrated into the most fertile lands, grown rice (a high status food) and hold in disdain 'wild' foods located at the extreme opposite end of the nature-culture continuum. The consumption of 'wild' foods is widely recognized as a symbol of poverty and low status by high-caste Hindus, policy makers and Tibeto-Burman groups desiring to raise their position in the Hindu caste system.

Milton (1996:167) pointed out that because "discourses generate diverse ways of understanding the world," they are said "to 'compete' in given social contexts." The term discourse is used to mean both speaking (involving symbols and meaning) and action (involving material transformation of society and environment) (Blaikie 1995:207). Policy makers, especially high-caste Hindus, have been affected by the labels they use and their views have resulted in policies seeking to control use of the *jangal*. This has often had a negative impact on those who are most dependent on it, that is, the foraging farmers. Many outsiders view the *jangal* as low status, peripheral and used and inhabited by "tribal" people. It is not only the visible resources but also the "hidden" resources that are important. Thus a political-economic arena exists where various actors have unequal access to power and this can have implications for environmental management.

The Rai's retention of social and religious institutions, especially those associated with the *jangal*, have helped them to resist outside influences and maintain their ethnic identity, in contrast to Buddhist Sherpas and Hindus. One important example is their resistance to Sanskritization which has had and can have a negative impact on the environment because of its socio-religious practices. Sanskritization beliefs are evident among economically well-off Sherpas who show a disdain for *jangal* edibles such as stinging nettles when guests (foreign and Brāhman) are

---

43 Hindus, who are poverty-stricken may not be able to follow the requirements of the dominate Hindu ideology (Berreman 1993:367).

present. In her studies in Kabre, Sacherer (1979b) reported on the adverse nutritional consequences of Sanskritization. She pointed out that whilst wealth may increase as one ascends the caste ladder, health, or at least nutrition, may decrease because along with upward caste, mobility restrictions on the diet increase. Sacherer referred to this phenomenon as wealth without health (see Chapter 6 for the example of how high-caste Hindu children complained of eye problems). Whilst for the Rais, health without wealth fits them as far as micro-nutrients are concerned. Rais have resisted this devaluation of the important food source from the *jaṅgal*. The status of the *sisnu* held by the Rai farmers is exemplified by their question to me, "Will we still be able to collect *sisnu* once the National Park is established?" Thomas-Slayter and Bhatt (1994:480) report that farmers in Ghusel Village Development Committee (Lalitpur District) plant *sisnu* (stinging nettles) "in abundance" because when they are available to cattle they directly improve the amount of fat and quantity of milk produced. Secondly, they are scarce in government forests, which indicates how much they are gathered and how important they are to the subsistent economy.

As well as the idea of pollution, another reason *jaṅgal* ingestibles have remained "hidden" from outsiders is that many of the ingestibles also contain toxins which protect the plants and render them inedible. However, the local users have cultural processing techniques to detoxify them, although they are often time consuming.

In this area, where *anikāl* (hunger season) is a recurrent problem, the farmer uses a poly-crop of biodiversity to meet his or her needs for livelihood. Although 'wild' foods grow in both agricultural fields and *jaṅgal* areas and are important in the diet, they have often been dismissed in food surveys and omitted from program planning (see Koppert 1986). Only if the researcher has a specific purpose to seek them out, are 'wild' foods noted, or are the locals paid attention to.

The Rais gather 'wild' greens. High caste people consider 'wild' greens "polluted," and some Western scientists dismiss them as "weeds." 'Wild' mushrooms are often considered desirable foods in many parts of the world. But for the Rais, mushrooms are less important than the "weeds" dismissed by other groups. Meilleur (1994:265) cautioned that "other, less narrowly 'anthropo-selfish' arguments must be found to increase concern for the plight of species having no apparent utility." This is because "funding is biased toward conservation programs involving species for which economic potential can be demonstrated" (Plotkin 1988).

Differing Cultural Views of Degradation

The meaning and implications of environmental degradation vary at the different management levels. The term "degradation" is cultural and needs to be defined from both the scientific and local perspective. Scientist use the term "degradation" to refer to a decrease in biological diversity, that is, change from an assumed pristine state to a relatively disturbed agricultural environment. Policy planners and national parks focus on the preservation of biodiversity and its promotion as a tourist attraction. For the local folks residing on the fringes, a certain amount of "degradation" is a natural process which in itself stimulates biodiversity. I challenge the assumption of a degrading forest based on qualitative data. Of course there is a certain minimal degradation from the daily wear and tear of forest disturbance; however, the problem arises when the threshold for natural regenerational capacities of the environment is exceeded (Sponsel 1992).

Disturbance in the forest acts to increase biodiversity for the rural farmer and creates a variety of resources in various niches. Swidden fields, for example, are less susceptible to pest infestation than permanent fields. The co-evolution of *chiraito* and swidden is another example. *Chiraito* collection and swidden agriculture are mutually enhancing, the former providing a source of local traditional medicine and now a source of income, and the latter, food supplies and cultivated crops. An added benefit of swidden fields is the presence of *jangal* ingestibles which appear simultaneously with the planting of the crops and continue into the fallow cycle. The Rai and the poorer Sherpa farmer do not want to degrade what is of utmost importance for their survival.

The Changing Boundaries of Land Use

We need to look at how the interplay between the various levels of management affects the use and management of the natural resources. Changing environmental, social and political conditions affect the system of property rights and land tenure. The Rai and Sherpa communities act differently with respect to the various resources and their dependence on them. The boundaries in the *jangal* are elastic, changing with the risk and uncertainty the farmers face and the demands and needs of the various resources. The farmers have learned to adapt to these changing boundaries. I have argued that the people are aware of environmental problems in their area. They plant fodder trees on their private land, and utilize indigenous non-formal user groups to manage the extraction of resources, such as *malingo* (*Arundinaria aristatla*). The landscape contains a mosaic of biodiversity created in part by patches in different stages of ecological succession. The farmers have evolved strategies expressed through their cultural models of how best to protect and use these resources because they are concerned for their children's and grandchildren's future.

There have been changes in government rules of land tenure in the last century in most areas of Nepal. But in my research area and that of the Limbus, the *kipat* system of common property had, in practice, been retained during my research period The harvesting of 'wild' resources from common property areas, such as *chiraito* was still the same. Although previously *chiraito* could be harvested by anyone even on private land, recently the patterns of access rights have changed because the market price of *chiraito* has increased creating a high demand for it. Now when it grows on private property it is considered to be privately owned.

At the level of the governmental agencies, boundaries have been set which will not allow the flexibility the people once had and still need. These boundaries represent the first time there has been external control on land use. Previously such use was loosely monitored by the *kipat* system. Thus the repercussions felt by the farmers are quite different now because they realize that they may not have the option of buffering the environmental insults which may occur.

The local situation is quite different from the cultural views held by higher levels of management. The Cadastral Survey is an example of how cultural beliefs held at different levels of management led to the degradation of the environment because there was no two-way flow of information. The farmers were allowing the swidden cultivated fields to regenerate back to forest so the soil would become fertile again. However, because they feared the government would take away swiddened land, the rotation of the swidden fallow cycle was shortened so the fields appeared to be under permanent cultivation, resulting in devastating forest destruction.

How Will Modernity Affect the Rais?

Outward migration is occurring more and more, and the recent trend is for younger household members to run away to Kathmandu, the capital, feeling the "bright lights" will be better than the drudgery of the farmwork they are faced with at home. Or perhaps they wish to escape an arranged marriage. The flow of laborers out of the area means there are not enough hands to weed the swidden gardens which are increasingly being abandoned.

As mentioned in the text, the Koshi Hills Rural Development Programme's major finding is that a household's access to extension services and credit is determined largely by the wealth and social status of the household members (Cassels et al 1987). These services – low interest credit for small farmers, agricultural extension coverage, livestock services and cottage industries – are not reaching the Rais and Sherpas in this area because of their remoteness.

This rural area has been a classic example of a subsistence economy relying mainly on bartering with neighboring groups for subsistence needs and minimal market interaction. The flow of cash has been minimal but now with the influx of tourists, and exposure to modernity through trekking, the situation is changing. Now

]the need for cash is growing. The Sherpas, being conveniently situated to be active in trekking, have benefited by modern changes. The appearance of substantial houses, radios, and Western clothing attest to this.

The Rais have limited contact with the market. The trekking routes by-pass their villages. The main source of cash is *chiraito* which has become an important medicinal export to India. Culturally they prefer to maintain their close ties with the *jangal* and are the most renowned shamans in the area. They retain their own language to a greater degree than most Tibeto-Burman groups and are suspicious of outsiders. Despite some outmigration of young people to Kathmandu, they fear that development will bring changes over which they have no control and so are ambivalent about modernization. Better communications with the government are needed, for example, regarding the management of *chiraito* (see Recommendations section).

The Gender Perspective

Poor households, and particularly women, have depended on the gathering of *jangal* resources on common property areas as a means of providing household food security for their families. *Chiraito*, formerly so abundant that it could even be gathered on private property, is now restricted so that outsiders can no longer have access since it has become the major source of cash. Also, the restriction on swidden cultivation in forested areas which are government land (formerly regarded as common property), further limits the wild growth of *chiraito* and the cultivation of food crops. Carpenter's (1991) Pakistani study showed that a decline in common land limits women's access to undomesticated sources of food [and the nutrients provided by these foods]. For poorer women with inadequate landholdings, these changes in land tenure may entail the loss of an important source of income which can cause dire nutritional and health-related problems within their families. Regarding natural resource management planning, women need to be included in all stages. McGuire (1993) pointed out that when women were encouraged to manage their forests, they were able to protect the 'wild' foods on which they depended.

There is a complex interaction between forestry and farming practices, and each community adopts distinctive adaptive strategies. There are gender, social class, and ethnic differences in *jangal* resource knowledge, use and management. A number of variables, including subsistence labor, seasonal outmigration of men, accessibility of the resources, and market demand, account for these differences.

Among the Rais and Sherpas, forestry, nutrition and health are linked together through women's use of *jangal* resources for household nutrition security (see Chapter 7). Rai and Sherpa women have developed various behavioral strategies (which vary according to their economic status and indigenous knowledge of the *jaṅgal*) in order to alleviate seasonal shortages in their food supply. Rai women use *jaṅgal* ingestibles as a regular supplement in their diet, while most Sherpa women gather them seasonally for variety and taste. Even when household food security needs are being met, social factors prevent an equal allocation of food within the household, with women in general getting a lesser share and, especially the daughter-in-law who is last in line. Women already bear a heavy proportion of the labor used for food production and processing. If the men are drawn off to wage labor, the women's burden will be even heavier. Furthermore, men often spend a substantial proportion of their wages on luxuries. The majority of women in my area do not speak Nepali, the *lingua franca* of Nepal, and the language often used in trade transactions. As an increasingly larger number of men migrate for longer periods of time, there may be a greater burden placed on women who will need to rely more on loans which currently have a high interest rate. Possible new approaches to loans will be discussed in the recommendation section.

Another negative consequence (besides nutrition and health) which may occur among some of the Sherpas in my research area is that of Sanskritization. Tibeto-Burman women have the greatest autonomy and equality with men as shown by studies among both Tibeto-Burman and Indo-Aryan groups throughout Nepal (Acharya and Bennett 1981). However, Berreman (1993:370) asserts that in India, Sanskritization initiates or deepens female subordination and is inherently counterproductive for women..." while on the other hand secular mobility, personal income and education are the "most promising mechanism for status enhancement of females." The secular mechanisms are promising because mobility exposes them to ideas, and income and education enhances their decision-making power and status. A few Sherpa women (both married and unmarried) have not been following the traditional rules and have gone trekking and put the money they earned aside in their own purses. At the moment, I have not seen any major changes among the Tibeto-Burmans in my area in following other Hindu rules which can adversely affect their status. However, this may change with increasing modernity.

What Do the Local People Perceive as Their Needs? And How Can They Be Met?

Planning cannot be done *a priori*. Local people must first be consulted. Rai and Sherpa women both cited inadequate health care and insufficient food supplies as their most pressing problems. Their concerns are expressed by this quote from a Sherpa woman who lives in the remote village of Saisima:

> *A Khaddyana (a food depot or storage place) is needed where we can purchase, or borrow on credit, staple foods needed for anikāl. During barkhā (the monsoon period), heavy rainfall*

causes landslides and bridges are unsafe to cross, making bartering trips to neighboring villages for food purchases very dangerous, if not impossible! Our children are dying, there is no doctor here, and we would like to have a health center.

Installation of a *khaddyana* is a short-term emergency measure which can prevent starvation, but long term solutions are needed. Women can significantly improve their own nutritional status and that of their children even without a major increase in agricultural labor or the total amounts of food available. The reduction of post-harvest food losses through new techniques to improve storage could make a major contribution to nutritional security (Zeitlin and Brown 1992).

One Rai farmer expressed the main concerns they face and how they feel development can help their lot.

Development is important. In order for us farmers to do business, we need money. If we take out a loan from rich people in the village, it takes a lot of money to repay it because the interest rate is very high. Income earned by us from selling alcohol or making bamboo baskets is very small and not enough to repay the loans which become bigger and bigger from the interest. If the government opens some kind of small industry that we can be involved in and receive a fair share, then I feel this program can help us a lot.

Recommendations

The following recommendations are based on the problems identified by the Rai and Sherpa farmers and their suggestions of how they can best be assisted in the development of community-based programs.

1. Co-management of swidden and village-based *chiraito* processing enterprises

Since the research was finished the Cadastral Survey has been completed. This means that the government has mapped and surveyed the land and so the *kipat* system has been abolished and these lands have been brought under government control. In Chapter 8, evidence was put forward which challenges the assumption that swidden cultivation when properly done is destructive to the environment. For example, Kleinman et al (1995:245) say that "the arguments against the viability of slash-and-burn agriculture as an ecologically sustainable form of crop production have pointed to the degradative effects on soil productivity. These arguments, however, do not take into account the great diversity in slash-and-burn systems. With proper management, swidden soil degradation is minimal." See also Schmidt-Vogt (1999). Since swidden cultivation can be a sustainable agricultural system it would be valuable if co-management of property rights between the government and local land users were established (Pinkerton 1987). This would serve a dual purpose to allow for the continuation of controlled swidden cultivation in the *jangal* and let the poorest Rais continue to use the *jangal* for survival. Furthermore *chiraito,* an income source for all farmers, flourishes in fallow swidden fields. It is the most important source of cash for Rais, and for Sherpas who do not trek, and the loss of this income would be disastrous for the financially weakest groups as they use the cash for food and other basic necessities.

It would be useful for the national park to encourage a local non-governmental organization to work with the villagers to set up some sort of coop to process and market the *chiraito* (see Chipeta 1995; Le Cup 1994). If the maximum benefit from the park for the environment and the people is to be gained, communication between the park personnel and the local population is essential, as problems can be discussed before they become serious and solutions can be found. For example, getting the co-operation of the local population is the best way to guard against undesirable expansion of swidden areas in to the *jangal*.

2. Low interest loans (based on the Grameen Bank)
In this area, indebtedness is a major problem and loans are taken out at exorbitant interest rates. The farmers are never able to repay the loans and are caught in a cycle of poverty. Micro-enterprises, such as the Grameen Bank, may help provide the means by which the farmers, particularly women, can improve their financial situation and status. The Grameen Bank was founded in 1976 in Bangladesh, to extend credit to poor landless villagers (persons owning less than 0.5 acres of cultivable land) who had no collateral to offer for loans. This micro-credit program promotes "peer-group" lending because five people of similar socio-economic status must agree to assume collective responsibility for each member's loans, a form of "social collateral." The repayment rate of these loans is 97 percent, and 94 percent of the total members are women. Thus the credit provides home-based self-employment options which act as a powerful tool to enhance the status of women and increase their economic decision-making power within the families. The beneficiaries of their income are their children, as shown by the improved nutritional status of children of Grameen Bank members, compared to non-Grameen Bank members (Bornstein 1996; Quanine 1990; Lawoti 1997).

3. Community Extension Services
Community extension services are especially important for farmers, although the remoteness of this area is a constraining factor. However, by sending several farmers, both male and female, to be trained in appropriate technology, they can then return and teach the other farmers. For example, post-harvest technology, with emphasis on homestead storage, preservation and processing, can reduce food loss and make more food available during *anikāl*.

The role of 'minor' *jangal* resources which provide major proportions of essential vitamins and minerals must be recognized in local food systems and their continued use supported. Because *jaṅgal* ingestibles are often considered low status, their importance to the local population should be explained by the school system and agricultural extension workers. A seasonal calendar can be prepared by extension workers together with men and women farmers in order to 1) determine when periods of food scarcity occur, understand the coping strategies used by the local farmers and to explore ways in which the extension workers can enhance the systems with their modern scientific knowledge. Since women are the main collectors of *jaṅgal* resources they have the information which will make them useful as an "early warning system" for the status of all of the resources, ingestible and non-ingestible.

Community forestry programs cannot work in isolation, but need to link up with other institutions and organizations in areas of health, agriculture, veterinary services and other income-generating activities. Community forestry initiatives need to incorporate nutritional and household food security concerns into their current programs. The people's use and knowledge of *jaṅgal* resources, along with the relationship between these resources and food security, should be documented.

Traditional and Western medical systems can work together. Many medicinal plants are available to treat various ailments. This folk medicine knowledge is in danger of being lost and needs to be documented and promoted. Women have a vital role regarding *jaṅgal* resources, but usually are not included in any major decision-making outside the household. This matter needs to be addressed by community foresters.

Most important is for foresters and other field workers to work together with the villagers in assessing their needs. This can ensure 'a communication channel in which information about technology and research needs and priorities flows with equal ease in both directions' (Raintree 1986). Only by addressing the social issues identified by the local people themselves and improving their socio-economic conditions, can conservation projects become self-sustaining.

References

Acharya, Meena and Lynn Bennett. 1981. *The Status of Women in Nepal*. Centre for Economic Development and Administration, Tribhuvan University, Kathmandu, Nepal.

Acheson, James M. 1989. Management of common-property resources. In: *Economic Anthropology*. Stuart Plattner, ed. Pp. 351-377. Stanford: Stanford University Press.

Adhikari, Ramesh K. and Miriam Krantz. 1989. *Child Nutrition and Health*. Kathmandu: Jeewan Printing Support Press.

Agarwal, Bina. 1991. *The Gender and Environment Debate. Lessons from India*. Presented at the 'Common Property, Collective Action and Ecology.' Conference held at Bangalore: Center for Environmental Sciences, Indian Institute of Science. August 19-21, 1991.

_____. 1986. *Cold Hearths and Barren Slopes. The Fuelwood Crisis In The Third World*. Maryland: The Riverdale Company, Inc.

_____. 1985. Women and technological change in agriculture: the Asian and African experience. In: *Technology and Rural Women*. Iftikhar Ahmed, ed. Pp. 67-114. London: George Allen and Unwin Publishers Ltd.

Agrawal, Arun. 1995. Dismantling the divide between indigenous and scientific knowledge. *Development and Change* 26:413-439.

Ahmed, Saleem. 1994. Meeting future food needs: the potential role of the potato in South Asia. *Outlook on Agriculture* 23(3): 177-182.

Alcorn, Janis B. 1996. Forest Use and Ownership: Patterns, Issues and Recommendations. In: *Forest Patches in Tropical Landscapes*. John Schelhas and Russell Greenberg, eds. Pp. 233-257. Washington, DC: Island Press.

_____. 1995. Ethnobotanical Knowledge Systems. A Resource for Meeting Rural Development Goals. In: *The Cultural Dimension of Developing Indigenous Knowledge Systems*. D. Michael Warren, L. Jan Slikkerveer and David Brokensha, eds. Pp. 1-12. London: LTD.

Alcorn, Janis B. and Augusta Molnar. 1996. Deforestation and human-forest relationships: what can we learn from India? In: *Tropical Deforestation. The Human Dimension*. Leslie E. Sponsel, Thomas N. Headland and Robert C. Bailey, eds. Pp. 99-121. New York: Columbia University Press.

Allen, Nicholas J. 1976a. Studies in the Myths and Oral Traditions of the Thulung Rai of East Nepal. Ph.D. dissertation. University of Oxford.

_____. 1976b. Approaches to illness in the Nepalese hills. In: *Social Anthropology and Medicine*. J. B. Loudon, ed. Pp. 500-552. London: Academic Press Inc.

_____. 1976c. Shamanism among the Thulung Rai. In: *Spirit Possession in the Nepal Himalayas*. John T. Hitchcock and Rex L. Jones, eds. Pp. 124-140. New Delhi: Vikas Publishing House Pvt Ltd.

Altman, Nathaniel. 1994. *Sacred Trees*. San Francisco: Sierra Club Books.

Amatya, S.M. and S. M. Newman. 1993. Agroforestry in Nepal: research and practice. *Agroforestry Systems* 21:215-222.

Amatya, Soorya Lal. 1988. Rural Development Through Decentralized Planning: A Case Study of Nepal. Working Paper #17. Honolulu: East-West Center.

Anderson, E.N. 1996. *Ecologies of the Heart. Emotion, Belief, and the Environment*. New York: Oxford University Press.

Anderson, Robert and Edna M. Mitchell. 1984. Children's health and play in rural Nepal. *Social Science and Medicine* 19(7):735-740.

Appasamy, Paul P. 1993. Role of non-timber forest products in a subsistence economy: the case of a joint forestry project in India. *Economic Botany* 47(3):258-267.

Applegate, G.B. and D.A. Gilmour. 1987. Operational Experiences in Forest Management Development in the Hills of Nepal. ICIMOD Occasional Paper No. 6. Kathmandu, Nepal.

APROSC. 1991. *Glossary of Some Important Plants and Animals Names in Nepal.* Agricultural Projects Services Centre. Kathmandu: Malla Press.

Aris, Michael. 1990. Man and nature in the Buddhist Himalayas. In: *Himalayan Environment and Culture.* N.K. Rustomji and Charles Ramble, eds. Pp. 85-101. Shimla: Indian Institute of Advanced Study.

Arnold, J.E.M. and J. Gabriel Campbell, 1986. Collective Management of Hill Forests in Nepal: The Community Forest Development Project. Proceedings of the International Conference on Common Property Resource Management, Annapolis, Maryland, 21-27 April, 1985. Washington, D.C.: National Academy of Sciences.

Aryal, Manisha. 1993. Diverted Wealth. The Trade in Himalayan Herbs. *HIMAL:* 9-18.

Ashby, Jacqueline A. and Douglas Pachico. 1987. Agricultural ecology of the mid-hills of Nepal. In: *Comparative Farming Systems.* B.L. Turner II and Stephen B. Brush, eds. Pp. 195-222. New York: The Guilford Press.

Baidya, Yogesh. 1991. Carotene Content of Some Prominent Food Plants of Nepal. *Nutrition Bulletin* 4(6):34-41. Joint Nutrition Support Programme. Lalitpur.

Bajracharya, Deepak. 1983. Deforestation in the food/fuel context. historical and political perspectives from Nepal. *Mountain Research and Development* 3(3):227-240.

_____. 1980. Fuelwood and Food Needs Versus Deforestation: An Energy Study of a Hill Village Panchāyat in Eastern Nepal. Energy for Rural Development Program Report Pr-80-2, Honolulu: East-West Center.

Baker, Paul T.A. 1973. Human adaptation to high altitude. In: *Man in Evolutionary Perspective.* C. Loring Brace and James Metress, eds. Pp. 411-425. New York: John Wiley and Sons.

Bangham, C.R. and J.M. Sacherer. 1980. Fertility of Nepalese Sherpas at moderate altitudes: comparison with high-altitude data. *Annals of Human Biology* 7(4):323-330.

Barlett, Peggy F. 1989. Introduction: dimensions and dilemmas of householding. In: *The Household Economy. Reconsidering the Domestic Mode of Production.* Richard R. Wilk, ed. Pp. 3-10. Boulder: Westview Press.

_____. 1982. *Agricultural Choice and Change: Decision Making in a Costa Rican Community.* New Brunswick, NJ: Rutgers University Press.

_____. 1980. *Agricultural Decision Making. Anthropological Contributions to Rural Development.* New York: Academic Press.

Basu, Amitabha, Roy, Subrata K., Mukhopadhyay, Barun, Bharati, Premananda, Gupta, Ranjan and Partha P. Majumder. 1986. Sex bias in intrahousehold food distribution: roles of ethnicity and socioeconomic characteristics. *Current Anthropology* 27(5):536-539.

Bates, Daniel G. and Fred Plog. 1991. *Human Adaptive Strategies.* New York: McGraw-Hill, Inc.

Beall, Cynthia M. 1983. Ages at menopause and menarche in a high-altitude Himalayan population. *Annals of Human Biology* 10(4):365-370.

Bennett, John W. 1990. Ecosystems, environmentalism, resource conservation, and anthropological research. In *The Ecosystem Approach in Anthropology. From Concept to Practice.* Emilio F. Moran, ed. Pp. 435-457. Ann Arbor: The University of Michigan Press.

_____. 1976. *The Ecological Transition: Cultural Anthropology and Human Adaptation.* New York: Pergamon Press.

_____. 1946 An interpretation of the scope and implications of social scientific research in human subsistence. *American Anthropologist* 48:553-573.

Bennett, Lynn. 1988. The role of women in income production and intra-household allocation of resources as a determinant of child nutrition and health. *Food and Nutrition Bulletin* 10(3):16-26.

Bernbaun, Edwin. 1992. *Sacred Mountains of the World.* San Francisco: Sierra Club Books.

_____. 1980. *The Way to Shambhala: A Search for the Mythical Kingdom Beyond the Himalayas.* Garden City, NY: Anchor Press.

Berreman, Gerald D. 1993. Sanskritization as female oppression in India. In: *Sex and Gender Hierarchies.* Barbara D. Miller, ed. Pp. 366-392. Cambridge: Cambridge University Press.

Berry, Sara S. 1980. Decision making and policy making in rural development. In: *Agricultural Decision Making. Anthropological Contributions to Rural Development.* Peggy F. Barlett, ed. Pp. 321-335. New York: Academic Press.

Bhadra, Binayak, Sharma, Sridhar, Khanal, Narendra, Joshi, Amrit and Bharat Sharma. 1991. *Management of Natural Resources: Arun III: Management of Basinwide Environmental Impacts Study.* Kathmandu: King Mahendra Trust.

Bhattarai, N.K. 1989 Ethnobotanical studies in Central Nepal: the ceremonial plant-foods. *Contributions to Nepalese Studies* 16:35-41.

Bista, Dor Bahadur. 1967. *People of Nepal.* Ministry of Information and Broadcasting, His Majesty's Government, Kathmandu, Nepal.

Bjonness, Inger-M. 1986. Mountain hazard perception and risk-avoiding strategies among the Sherpas of Khumbu Himal, Nepal. *Mountain Research and Development* 6(4):277-292.

Blaikie, Piers. 1995. Changing environments or changing views? A political ecology for developing countries. *Geography* 80(3):203-214.

Blaikie, P., Cameron J. and D. Seddon. 1980. *Nepal in Crisis.* New Delhi: Oxford University Press.

Bornstein, David. 1996. The Grameen Bank: Loans for poor people with "social collateral." *VIEWPOINT.* Winter. Pp. 14-15.

Boserup, Ester. 1965. *The Conditions of Agricultural Growth. The Economics of Agrarian Change Under Population Pressure.* New York: Aldine Publishing Company.

Bremness, Lesley. 1994. *Herbs. Eyewitness Handbooks.* New York: DK. Publishing, Inc.

Brokensha, David W., Warren, D. M. and Oswald Werner. 1980. *Indigenous Knowledge Systems and Development.* Lanham, MD: University Press of America.

Brower, Barbara. 1991. *Sherpa of Khumbu. People, Livestock and Landscape.* Delhi: Oxford University Press.

Bunting, Bruce W., Sherpa, Mingma N. and Michael Wright. 1991. Annapurna Conservation Area: Nepal's new approach to protected area management. In: *Resident Peoples and National Parks. Social Dilemmas and Strategies in International Conservation.* Patrick C. West and Steven R. Brechin, eds. Pp. 160-172. Tucson: The University of Arizona Press.

Burbage, M.B. 1981. *Report on a visit to Nepal: the medicinal plant trade in the KHARDEP area: a study of the development potential.* London: Tropical Products Institute.

Burton, Benjamin T. 1965. *The Heinz Handbook of Nutrition.* New York: McGraw-Hill.

Burton, Michael L., Schoepfle G. Mark and Marc L. Miller. 1986. Commentary. natural resource anthropology. *Human Organization* 45(3):261-269.

Byers, Alton C. 1987. An assessment of landscape change in the Khumbu region of Nepal using repeat photography. *Mountain Research and Development* 7:1(77-81).

_____. 1986. A geomorphic study of man-induced soil erosion in the Sagarmatha (Mt. Everest) National Park, Khumbu, Nepal: report of activities of the UNU/MAB (Nepal) mountain hazards mapping project, phase II. *Mountain Research and Development.* 6(1):83-87.

Byrne, Peter C. 1994. Letter dated January 12th, 1994. Sent from Mount Hood, Oregon.

Cadeliña, Rowe V. 1985. *In Time of Want and Plenty, The Batak Experience*. Siliman University, Dumaquete City, Philippines.

Caldwell, M.J. and I.C. Enoch. 1972. Riboflavin content of Malaysian leaf vegetables. *Ecology, Food and Nutrition* 1:301-312.

Caplan, Lionel. 1970. *Land and Social Change in East Nepal. A Study of Hindu-Tribal Relations*. Berkeley: University of California Press.

Carpenter, Carol. 1991. Women's Livestock Production and Natural Resources in Pakistan. *Environment and Policy Institute Working Paper No. 26*, Honolulu: East-West Center.

Carpenter, Chris and Robert Zomer. 1996. Forest ecology of the Makalu-Barun National Park and Conservation Area, Nepal. *Mountain Research and Development* 16(2):135-148.

Carson, B. 1985. Erosion and sedimentation processes in the Nepalese Himalayas. *ICIMOD Occasional Paper No. 1*, Kathmandu.

Cashdan, Elizabeth. 1990. *Risk and Uncertainty in Tribal and Peasant Economies*. Boulder, Colorado: Westview Press.

Casimir, M.J. 1988. Nutrition and socio-economic strategies in mobile pastoral societies in the Middle East with special reference to West Afghan Pashtuns. In: *Coping With Uncertainty in Food Supply*. I. de Garine and G.A. Harrison, eds. Pp. 337-359. Oxford: Clarendon Press.

Cassels, Claudia, Wijga, Alet, Pant, Mahesh and Nabarro, David. 1987. Coping strategies of East Nepal farmers: can development initiatives help? *Five year study of the impact of the Kosi Hill area rural development programme*. HMG and UKODA, Kathmandu.

CBS. 1996. Nepal Living Standards Survey 1995-96. Statistical Report. Main Findings. Volume I. CBS/ HMG Nepal.

_____. 1987. Statistic Year Book of Nepal, Central Bureau of Statistics, Kathmandu.

_____. 1985. National Sample Census of Agriculture. 1981/2. Kathmandu: Central Bureau of Statistics.

Cernea, Michael M. 1992. The privatization of the commons: Land tenure and social forestry development in Azad Kashmir. In: *Sociology of Natural Resources in Pakistan and Adjoining Countries*. Michael R. Dove and Carol Carpenter, eds. Pp. 188-217. Lahore: Vanguard Books Pvt. Ltd.

Chambers, Robert. 1979. Editorial. *International Development Studies Bulletin* 10(2).

Chevallier, Andrew. 1996. *The Encyclopedia of Medicinal Plants*. New York: DK Publishing Inc.

Chhetri, R.B. and T.R. Pandey. 1992. User Group Forestry in the Far-Western Region of Nepal. Case studies from Baitadi and Accham. Kathmandu: ICIMOD.

Chipeta, M.E. 1995. Making of non-wood forest products programmes succeed: Lessons from small-scale forest-based enterprises. In: *Report of the International Expert Consultation on Non-Wood Forest Products*. Pp. 147-155. Yogyakarta, Indonesia. 17-27 January 1995. Rome: FAO.

Collins, John J. 1978. *Primitive Religion*. Totowa, N.J.: Rowman and Littlefield.

Conklin, Harold C. 1954. An ethnoecological approach to shifting agriculture. Transactions of the New York academy sciences 17:133-142. Reprinted 1969 in: *Environment and Cultural Behavior. Ecological Studies in Cultural Anthropology*. A.P. Vayda, ed. New York: The Natural History Press.

Conlin, S. and A. Falk. 1979. A study of the socio-economy of the Koshi Hill area: Guidelines for planning an integrated rural development program. *KHARDEP Report No. 3*. Dhankuta, Nepal.

Cornu, Philippe. 1990. *L'astrologie Tibétaine*. Collection Présences.

Coursey, D. G. 1967. *Yams*. London: Longmans.

Cronin, Edward W. 1979. *The Arun: A Natural History of the World's Deepest Valley.* Boston: Houghton Mifflin Company.

Crosby, Jr., Alfred W. 1972. *The Columbian Exchange. Biological and Cultural Consequences of 1492.* Westport: Greenwood Press.

Daniggelis, Ephrosine K. 1996. Unpublished field report, Tibet.

_____. 1994a *Jangal* Resource Use: Adaptive Strategies of Rais and Sherpas in the Arun Valley of Eastern Nepal. In: *The Anthropology of Nepal, Peoples, Problems and Processes.* Michael Allen, ed. Pp. 49-63. Kathmandu: Mandala Book Point.

_____. 1994b The *Jangal's* 'Hidden' Wealth: Women's Behavioral Response to Food Scarcity in Eastern Nepal. *Multipurpose Tree Species Network Research Series Report Number 22.* Forestry/Fuelwood Research and Development Project (F/FRED), Winrock International and the U.S. Agency for International Development. Bangkok.

Dankelman, Irene and Joan Davidson. 1988. *Women and Environment in the Third World. Alliance for the Future.* London in Association with IUCN. Earthscan Publications Ltd.

Das, Sarat C. 1970. [1902] *Contributions to the Religion and History of Tibet.* New Delhi: Manjusri Publishing House.

Dawson, Penny. 1991. Personal Communication. Kathmandu, Nepal.

deBeer, Jenne H. and Melanie J. McDermott. 1989. *The Economic Value of Non-Timber Forest Products in Southeast Asia with Emphasis on Indonesia, Malaysia and Thailand.* Amsterdam: Netherlands Committee for IUCN.

De Garine, Igor and Sjors Koppert. 1990. Social adaptation to season and uncertainty in food supply. In: *Diet and Disease. In Traditional and Developing Societies.* G.A. Harrison and J. C. Waterlow, eds. Pp. 240-289. Cambridge: Cambridge University Press.

Denholm, Jeanette. 1991. Agroforestry in mountain areas of the Hindu Kush-Himalayan Region. ICIMOD Occasional Paper No. 17. Kathmandu: ICIMOD.

Dettwyler, Kathyrn G. 1989. Styles of infant feeding: parental/caretaker control of food consumption in young children. *American Anthropologist* 91:696-703.

Dewey, Kathyrn G. 1979. Commentary, agricultural development, diet and nutrition. *Ecology of Food and Nutrition* 8:265-273.

Diemberger, Hildegarde. 1992. *The Hidden Valley of the Artemisia.* Tibetan sacred geography as seen within a framework of representations of nature, society and cosmos. Ph.D. dissertation, University of Wien.

_____. 1988. Unpublished Field Report on Makalu-Barun to the Woodlands Mountain Institute. Kathmandu.

Donovan, Deanna G. 1981. Fuelwood: How Much Do We Need? Hanover, New Hampshire: Institute of Current World Affairs.

Dornstreich, M. 1977. The Ecological Description and Analysis of Tropical Subsistence Patterns: An Example from New Guinea. In: *Subsistence and Survival: Rural Ecology in the Pacific.* T. Bayliss-Smith and R.G. Feachem, eds. London: Academic Press.

Dove, Michael R. 1994. Marketing the Rainforest: 'Green' Panacea or Red Herring? Asia Pacific Issues. Analysis from the East-West Center No. 13, Honolulu.

_____. 1993. Smallholder rubber and swidden agriculture in Borneo: A sustainable adaptation to the ecology and economy of the tropical forest. *Ecnomic Botany* 47(2):136-147.

_____. 1992 The dialectical history of "jungle" in Pakistan: an examination of the relationship between nature and culture. *Journal Anthropology Research* 48:231-253.

_____. 1985. The agroecological mythology of the Javanese and the political economy of Indonesia. *Indonesia No. 39.*

_____. 1984. Man, land and game in Sumbawa: Some observations on agrarian ecology and development policy in Eastern Indonesia. *East-West Environment Policy Institute Reprint No. 82.*

_____. 1983. Theories of swidden agriculture and the political economy of ignorance. *Agroforestry Systems* 1:85-99.

Dove, Michael R. and Carol Carpenter. 1992. Introduction: The sociology of natural resources in Pakistan. In: *Sociology of Natural Resources in Pakistan and Adjoining Countries.* Michael R. Dove and Carol Carpenter, eds. Pp. 1-30. Lahore: Vanguard Books Pvt. Ltd.

Dove, Michael R. and Daniel M. Kammen. 1997. The epistemology of sustainable resource use: Managing forest products, swiddens and high-yielding variety crops. *Human Organization* 56(1): 91-101.

Dove, Michael R. and Abdul L. Rao. 1990. Common resource management in Pakistan: Garrett Hardin in the *Junglat. Environment Policy Institute Working Paper No. 23.* East-West Center: Honolulu.

Downs, Hughes R. 1980. *Rhythms of a Himalayan Village.* San Francisco: Harper and Row.

Downs, James F. 1964. Livestock production and social mobility in high altitude Tibet. *American Anthropologist* 66(5):115-119.

Dunn, F.L. 1975. Rain-Forest Collectors and Traders. A Study of Resource Utilization in Modern and Ancient Malaya. Monographs of the Malaysian Branch Royal Asiatic Society No. 5. Kuala Lumpur.

Dunsmore, J.R. 1988. Mountain environmental management in the Arun River basin of Nepal. *ICIMOD Occasional Paper No. 9.* Kathmandu.

Durrenberger, E. Paul. 1996. Economic anthropology. In: *Encyclopedia of Cultural Anthropology. Volume 2.* David Levinson and Melvin Ember, eds. New York: Henry Holt and Company.

Dyhrenfurth, N.G. 1959. Slick-Johnson Nepal snowman expedition. *The American Alpine Journal.* Pp. 324-326.

Eckersley, R. 1992. *Environmentalism and Political Theory: Toward an Ecocentric Approach.* London: University College London Press.

Eckholm, E.P. 1976. *Losing Ground: Environmental Stress and World Food Prospects.* Worldwatch Institute: New York.

Edgerton, Robert B. 1992. *Sick Societies. Challenging the Myth of Primitive Harmony.* New York: The Free Press.

Edwards, David M. 1993. The marketing of non-timber forest products from the Himalayas: the trade between East Nepal and India. *Rural Development Forestry Network Paper 15b.* London.

Ellen, Roy. 1986. *Environment, Subsistence and System: The Ecology of Small-Scale Social Formations.* Cambridge: Cambridge University Press.

Elwin, Verrier. 1994. Civilizing the savage. In: *Social Ecology.* Ramachandra Gudha, ed. Pp. 249-274. Delhi: Oxford University Press.

English, Richard. 1985. Himalayan state formation and the impact of British rule in the 19th Century. *Mountain Research and Development* 5(1): 61-78.

Eppsteiner, Fred, ed. 1988. *Path of Compassion. Writings on Socially Engaged Buddhism.* Berkeley: Parallax Press.

Escobar, Arturo. 1995. *Encountering Development. The Making and Unmaking of the Third World.* Princeton: Princeton University Press.

Etkin, Nina L. 1994. The cull of the wild. In: *Eating on the Wild Side. The Pharmacologic, Ecologic, and Social Implications of Using Noncultigens.* Nina L. Etkin, ed. Pp. 1-21. Tucson: The University of Arizona Press.

_____. 1993. Anthropological methods in ethnopharmacology. *Journal of Ethnopharmacology* 38:93-104.

Etkin, Nina L. and Paul J. Ross. 1991. Recasting malaria, medicine and meals: a perspective on disease adaptation. In: *The Anthropology of Medicine: From Culture to Medicine (2nd Ed.)*. L. Romanucci-Ross, D.E. Moerman and L.R. Tancredi, eds. New York: Bergin and Garvey.

Falconer, Julia. 1990. "Hungry season" food from the forests. UNASYLVA 41:14-19.

Falconer, Julia and J.E.M. Arnold. 1991. Household food security and forestry: An analysis of socio-economic issues. *Community Forestry Note #1*. Rome: FAO.

FAO. 1992. *Improving Household Food Security Theme Paper no. 1. Major Issues for Nutrition Strategies*. International Conference on Nutrition. Rome: Food and Agricultural Organization.

_____. 1989a. *Forestry and Nutrition*. A Reference Manual. Food and Agriculture Organization of the United Nations.

_____. 1989b *Forestry and Food Security*. FAO Forestry Paper 90. Food and Agriculture Organization of the United Nations.

Farquharson, Margaret S. 1976. Growth patterns and nutrition in Nepali children. *Archives of Disease in Childhood* 51:3-12.

Feeny, David, Berkes, Fikret, McCay, Bonnie J. and James M. Acheson. 1990. The tragedy of the commons: twenty-two years later. *Human Ecology* 18(1):1-19.

Fernandes, Walter and Geeta Menon. 1987. *Tribal Women and Forest Economy. Deforestation, Exploitation and Status Change*. New Delhi: Indian Social Institute.

Ferro-Luzzi, G. Eichinger. 1980. Food avoidance at puberty and menstruation in Thailand. an anthropological study. *Food, Ecology and Culture :93-100*.

Fisher, Robert J. 1990. Review Essay. The Himalayan dilemma: finding the human face. *Pacific Viewpoint* 31(1):69-76.

_____. 1989 Indigenous systems of common property forest management in Nepal. *Environment and Policy Institute Working Paper 18*. Honolulu: East-West Center.

_____. 1987. Confusing numbers with facts: A note of caution about the results of quantitative survey questionnaires. *Banko Janakari* 1(3).

Forbes, Ann. 1993. Disputing claims to *kipat*: The intersection between local and national systems of governance. South Asian Studies Conference. November 5-7, 1993. Madison, Wisconsin.

Fortmann, Louise and Diane Rocheleau. 1985 Women and agroforestry: four myths and three case studies. *Agroforestry Systems* 2:253-272.

Fox, Jeff. M. 1995, ed. *Society and non-timber forest products in Tropical Asia*. East-West Center Occasional Papers Environment Series No. 19, Honolulu.

_____. 1993. Forest Resources in a Nepali Village in 1980 and 1990: The Positive Influence of Population Growth. *East-West Center Reprints Environment Series No. 10*. Honolulu: East-West Center.

_____. 1984 Firewood consumption in a Nepali village. *Environmental Management* 8(3):243-250.

_____. 1983. Managing public lands in a subsistence economy: The perspective from a Nepali village. Ph.D. dissertation, University of Wisconsin, Madison.

Frake, Charles O. 1962. Cultural ecology and ethnography. *American Anthropologist* 64 (1,pt.1):53-59.

Freeman, J.D. 1955. *Iban Agriculture*. London: Her Majesty's Stationary Office.

French, Rebecca R. 1995. *The Golden Yoke. The Legal Cosmology of Buddhist Tibet*. Ithaca: Cornell University Press.

Fricke, Thomas. E. 1989. Introduction: human ecology in the Himalaya. *Human Ecology* 17(2):131-145.

_____. 1984. Talking about the household: a Tamang case from North Central Nepal. *Himalayan Research Bulletin* 4(2):17-30.

Furer-Haimendorf, Christoph von. 1975 *Himalayan Traders*. New York: St. Martin's Press.

_____. 1964. *The Sherpas of Nepal. Buddhist Highlanders*. Berkeley: University of California Press.

Gadgil, Madhav and V. D. Vartak. 1994. The sacred uses of nature. In: *Social Ecology*. Ramachandra Guha, ed. Pp. 82-89. Delhi: Oxford University Press.

Gaenszle, Martin. 1996. Raising the head-soul-a ritual text of the Mewahang Rai. *Journal of the Nepal Research Centre* 10:77-93.

_____. 1993 Ancestral types: mythology and the classification of deities among the Mewahang Rai. *Purusautha* 15:197-217.

_____. 1992. Poverty and Inequality: Traditional Self-Images Among Rai in the Western Arun Valley. Unpublished paper presented at Winrock International. Kathmandu.

Gaenszle, Martin, Bieri, Albin and Majan Garlinski. 1990. Deva and Cintā. Two Rituals Practised by the Rai of the Sankhuwa Valley in East Nepal. An Audio-Visual Documentation. University of Zurich.

Geertz, Clifford. 1965. Religion as a cultural system. In: *Anthropological Approaches to the Study of Religion*. Michael Banton, ed. Pp. 167-178. London: Tavistock Publications.

Ghirmire, Krishna. 1992. *Forest or Farm? The Politics of Poverty and Land Hunger in Nepal*. Delhi: Oxford University Press.

Gibbon, David, Joshi, Yagya R., Kumar, Sharan; Schultz, Michael; Thapa, Man B. and Mitra Pd. Upadhay. 1988. A study of the agricultural potential of Chheskam Panchāyat. *Parkribas Agricultural Center Technical Paper 95*. Kathmandu.

Gilmour, Donald A. and Robert J. Fisher. 1991. *Villagers, Forests and Foresters. The Philosophy, Process and Practice of Community Forestry in Nepal*. Kathmandu: Sahayogi Press.

Gilmour, D.A., King, G.C. and M. Hobley. 1989. Management of forests for local use in the hills of Nepal. I. Changing forest management paradigms. *Journal of World Forest Resource Management* 4:93-110.

Gilmour, Donald A. 1989. Forest Resources and Indigenous Management in Nepal. Working Paper No. 17. Honolulu: East-West Center.

_____. 1988. Not seeing the trees for the forest: A re-appraisal of the deforestation crisis in two hill districts of Nepal. *Mountain Research and Development* 8(4):343-350.

Gitteltsohn, Joel. 1991. Opening the box: Intrahousehold food allocation in rural Nepal. *Social Science and Medicine* 33(10):1141-1154.

Godoy, Ricardo, Brokaw, Nicholas and David Wilkie. 1995. The effect of income on the extraction of non-timber tropical forest products: model, hypotheses, and preliminary findings from the Sumu Indians of Nicaragua. *Human Ecology* 23(1):29-52.

Godoy, Ricardo and Ruben Lubowski. 1992. Guidelines for the economic valuation of nontimber tropical-forest products. *Current Anthropology* 33(4):423-433.

Goldstein, Melvyn C. and Cynthia M. Beall. 1990. *Nomads of Western Tibet. The Survival of a Way of Life*. Berkeley: University of California Press.

Golpaldas, Tara, Gupta, Anjali and Kalpna Saxena. 1983a. The impact of Sanskritization in a forest-dwelling tribe of Gujarat, India. ecology, food consumption patterns, nutrient intake, anthropometric, clinical and hematological status. *Ecology of Food and Nutrition* 12:217-227.

_____. 1983b. The phenomenon of Sanskritization in a forest-dwelling tribe of Gujarat, India. Nutrient intakes and practices in the special groups. *Ecology of Food and Nutrition* 13:1-8.

Gopalan, C., Rama Sastri, B.V. and S.C. Balasubramanian. 1984. *Nutritive Value of Indian Foods.* National Institute of Nutrition Indian Council of Medical Research, Hyderabad, India. Indian Council of Medical Research, CMR, New Delhi.

Griffin, D.M., Shepherd, K.R. and T.B.S. Mahat. 1988. Human impact on some forests of the middle hills of Nepal. part 5. comparisons, concepts and some policy implications. *Mountain Research and Development* 8(1):43-52.

Grivetti, Louis Evan. 1981. Cultural nutrition: anthropological and geographical themes. *Annual Review of Nutrition* 1:47-68.

Gross, Daniel R. and Barbara A. Underwood. 1971. Technological change and caloric costs: sisal agriculture in Northeastern Brazil. *American Anthropologist* 73:725-740.

Gubhaju, B.B. 1975. Fertility and mortality in Nepal. *Journal of Nepal Medical Association* 13:115-128.

Guillet, David. 1983. Toward a cultural ecology of mountains: The Central Andes and the Himalayas compared. *Current Anthropology* 24(5):561-574.

Gunatilleke, Iaun, Gunatilleke, C.V.S. and P. Abeygunawardena. 1993. Interdisciplinary research towards management of non-timber forest resources in lowland rain forests of Sri Lanka. *Economic Botany* 47(3):282-290.

Gunatilake, H.M., Senaratne, D.M.A.H. and P. Abeygunawardena. 1993. Role of non-timber forest products in the economy of peripheral communities of Knuckles national wilderness area of Sri Lanka: a farming systems approach. *Economic Botany* 47(3):275-281.

Gurung, Barun. 1995. A cultural approach to natural resource management: A case study from Eastern Nepal. In: *Beyond Timber: Social, Economic and Cultural Dimensions of Non-Wood Forest Products in Asia and the Pacific.* Patrick B. Durst and Ann Bishop, eds. Pp. 237-245. Bangkok: FAO

Gurung, Durga Kumari. 1987 Women's Participation in Forestry: A Case Study of Akrang Village. Forestry Research Paper Series Number 10. HMG-USAID-GTZ-IDRC-Ford Winrock Project.

Gurung, Om. 1994. Historical Dynamics of Resource Degradation in the Nepal Himalayas. In: *The Anthropology of Nepal, Peoples, Problems and Processes.* Michael Allen, ed. Pp. 82-96. Kathmandu: Mandala Book Point.

Gurung, Sumitra M. 1989. Human perception of mountain hazards in the Kakani-Kathmandu area: experiences from the middle mountains of Nepal. *Mountain Research and Development* 9(4):353-364.

Hammett, A.L. 1994. Developing community-based market information systems. In: *Marketing of Multipurpose Tree Products in Asia.* Raintree, J.B. and Francisco, H.A., eds. Pp. 289-300. Proceedings of an International Workshop held in Baguio City, Philippines, 6-9 December 1993. Winrock International, Bangkok.

HanBon, Gerd. 1991. *The Rai of Eastern Nepal: Ethnic and Linguistic Grouping. Findings of the Linguistic Survey of Nepal.* Werner Winter, ed. Linguistic Survey of Nepal and Centre for Nepal and Asian Studies, Tribhuvan University, Kathmandu.

Hardesty, D.L. 1977. *Ecological Anthropology.* New York: John Wiley & Sons.

Hardin, Garrett. 1968. The tragedy of the commons. *Science* 162:1243-1248.

Harris, Marvin. 1977. *Cannibals and Kings: The Origins of Cultures.* New York: Random House.

_____. 1965. The myth of the sacred cow. In: *Man, Culture and Animals: The Role of Animals in Human Ecological Adjustments.* A. Leeds and A.P. Vayda, eds. Washington: American Association for the Advancement of Science, Publication 78.

Hecht, S.B., Anderson, A.B. and P. May. 1988. The subsidy from nature: shifting cultivation, successional palm forests, and rural development. *Economic Botany* 47(1):25-35.

Hetzel, Basil S. 1990. The iodine deficiency disorders. In: *Diet and Disease in Traditional and Developing Societies*. G.A. Harrison and J.C. Waterlow, eds. Pp. 114-136. Cambridge: Cambridge University Press.

HMG, Nepal. 1988. *Master plan for the forestry sector, Nepal. Forest-based industries development plan-Part II*. His Majesty's Government of Nepal. FINNIDA/ADB/HMG, Kathmandu.

Hodgson, B.R. 1880. *Miscellaneous Essays Relating to Indian Subjects*. Volume I. London.

Hoskins, Marilyn. 1985. The promise in trees. *Food and Nutrition* 11:44-46.

_____. 1983 *Rural Women, Forest Outputs and Forestry Projects*. Rome: FAO.

Humphrey, Caroline. 1992. Fair dealing, just rewards: The ethics of barter in North-east Nepal. In: *Barter, Exchange and Value. An Anthropological Approach*. Caroline Humphrey and Stephen Hugh-Jones, eds. Cambridge: Cambridge University Press.

_____. 1980. Kosi Hill Area Rural Development Programme. Report on a Field-Study in North Sankhuwasabha Sept-Nov 1979. *KHARDEP Report No. 7*. KHARDEP, Kathmandu.

Huss-Ashmore, Rebecca and Susan L. Johnston. 1994. Wild plants as cultural adaptations to food stress. In: *Eating on the Wild Side. The Pharmacologic, Ecologic and Social Implications of Using Noncultigens*. Nina L. Etkin, ed. Pp. 62-82. Tucson: The University of Arizonia Press.

Ingels, Andrew W. 1995. Religious beliefs and rituals in Nepal. Their influence on forest conservation. In: *Conserving Biodiversity Outside Protected Areas. The Role of Traditional Agro-ecosystems*. Patricia Halladay and D.A. Gilmour, eds. Pp. 205-224. Gland, Switzerland: IUCN.

Inserra, A.E. 1989. Women's participation in community forestry in Nepal. *Banko Janakari* 2(2):119-120.

ITTO. 1991. *Sustainable Development of Tropical Forests*. 4th International Seminar. Held in conjunction with the 9th Council Session of the International Tropical Timber Organization in Japan, 17 November 1990.

Ives, Jack D. 1987. The theory of Himalayan environmental degradation: its validity and application challenged by recent research. *Mountain Research and Development* 7(3):189-199.

Ives, Jack D. and Bruno Messerli. 1989. *The Himalayan Dilemma. Reconciling Development and Conservation*. London and New York: Routledge.

Ives, Jack D., Messerli, Bruno and Michael Thompson. 1987. Research strategy for the Himalayan region conference conclusions and overview. *Mountain Research and Development* 7(3):332-344.

Iyer, L.A. Krishna. 1968. *The Social History of Kerala. Volume 1. The Pre-Davidians*. Madras: Book Centre Publishing.

Jackson, Cecile. 1993. Environmentalisms and gender interests in the Third World. *Development and Change* 24:649-677.

Jamieson, Neil L. 1984. *Multiple perceptions of environmental issues in rural Southeast Asia: The implications of cultural categories, beliefs, and values for environmental communication*. Working Paper, East-West Center, Honolulu.

Jerome, N.W., Kandel, R. F. and G. H. Pelto. 1980. *Nutritional Anthropology: Contemporary Approaches to Diet and Culture*. Pleasantville, New York: Redgrave Publishing Company.

Jodha, N.S. 1994. Common Property Resources and the Rural Poor. In: *Social Ecology*. Ramachandra Guha, ed. Pp. 150-187. Delhi: Oxford University Press.

Johnson, Allen. 1980. Ethnoecology and planting practices in a swidden agricultural system. In: *Indigenous Knowledge Systems and Development*. David W. Brokensha, D.M. Warren and Oswald Werner, eds. Pp. 49-66. Lanham: University Press of America.

Jones, Rex L. 1976. Limbu spirit possession and shamanism. In: *Spirit Possession in the Nepal Himalayas*. John T. Hitchcock & Rex L. Jones, eds. Pp. 29-55. New Delhi: Vikas Publishing House Pvt LTD.

Jones, Shirley K. 1976. Limbu spirit possession: a case study. In: *Spirit Possession in the Nepal Himalayas*. John T. Hitchcock & Rex L. Jones, eds. Pp. 22-28. New Delhi: Vikas Publishing House Pvt LTD.

Joshi, Amrit L. 1989. *Literature review resource management in community forestry*. Report for Centre for Economic Development and Administration, Tribuhuvan University, Kathmandu.

Joshi, Durga Datt. 1982. *Yak and Chauri Husbandry in Nepal*. Kathmandu, Nepal.

Joshi, N. 1993. Personal communication. Kathmandu, Nepal.

Kabilsingh, Chatsumarn. 1996. "Early Buddhist views on nature." In: *This Sacred Earth. Religion, Nature, Environment*. Roger S. Gottlieb, ed. New York: Routledge.

Karan, Pradyumna P. and Hiroshi Ishii. 1994. *Nepal. Development and Change in a Landlocked Himalayan Kingdom*. Monumenta Serindica No. 25. Institute for the Study of Languages and Cultures of Asia and Africa, Tokyo University of Foreign Studies.

Kawakita, J. 1984. *A Proposal for the Revitalization of Rural Areas based on Ecology and Participation through the Experiences of ATCHA in the Nepalese Hill Area*. Tokyo: Association for Technical Cooperation to the Himalaya Areas. Research Report No. R-84-2.

Khadka, Ramesh J. 1997. Personal communication, Honolulu, Hawaii.

Khadka, R.J., Gurung, B.D., Tiwari, T.P., Gurung, G.B., Ghimire, R.P., Adhikari, K.B., and D.N. Shrestha. 1994. A study on *chiraito* (*swertia chirata*): the high-altitude cash crop of the Eastern Hills of Nepal. *PAC Working Paper No. 107* Pakhribas Agricultural Centre, Kathmandu.

Khatri, Deep B. 1994. Nepal. In: *Non-Wood Forest Products in Asia*. Patrick Durst, Ward Ulrich and M. Kashio, eds. Pp. 73-79. RAPA Publication 28. RAPA, FAO. Bangkok.

King, G.C., Hobley, Mary and D.A. Gilmour. 1990. Management of Forests for Local Use in the Hills of Nepal. 2. Towards the Development of Participatory Forest Management. *Journal of World Forest Resource Management* 5:1-13.

Kleinman, P.J.A., Pimentel, D. and R.B. Bryant. 1995. The ecological sustainability of slash-and-burn agriculture. *Agriculture, Ecosystems and Environment* 52(2,3):235-249.

Kohn, Tamara. 1992. Guns and garlands: Cultural and linguistic migration through marriage. *Himalayan Research Bulletin* 12(1-2):27-33.

Koppert, G. 1986. Anthropology of Food and Nutrition in the Middle Hills of Nepal: A Preliminary Report on a Nutrition Survey in Salme, Nuwakot District, Nepal. In: *Nepal Himalaya: Geo-ecological Perspectives*. S.C. Joshi ed. Pp. 206-222. Naini Tal, India: Himalayan Research Group.

Krantz, Miriam. 1991. Personal Communication. United Nations Mission of Nepal.

Krause, Inga-Britt. 1988. Caste and labour relations in Northwest Nepal. *ETHNOS* 53(1-2):5-36.

Kuhnlein, Harriet V. and Nancy J. Turner. 1991. *Traditional Plant Foods of Canadian Indigenous Peoples: Nutrition, Botany and Use*. Philadelphia: Gordon and Breach Science Publishers.

Kumar, Shubh K. and David Hotchkiss. 1988 Consequences of Deforestation for Women's Time Allocation, Agricultural Production, and Nutrition in Hill Areas of Nepal. *International Food Policy Research Institute Research Report 69*.

Kunwar, Ramesh R. 1989. *Fire of Himal. An Anthropological Study of the Sherpas of Nepal Himalayan Region*. New Delhi: Nirala Publications.

Laderman, Carol. 1981. Symbolic and empirical reality: a new approach to the analysis of food avoidances. *American Ethnologist* 468-493.

Lampietti, Julian A. and John A. Dixon. 1995. To See the Forest for the Trees: A Guide to Non-Timber Forest Benefits. *Environment Department Paper No.013*. The World Bank.

Larrick, J. W. 1991. The methyl xanthine hypothesis: does tea consumption by Tibetan natives blunt the effects of high altitude? *Medical Hypotheses* 34(1):99-104.

Lawoti, Dovan. 1997. Correspondence received from Kathmandu, Nepal.

Lawoti, Mahendra. 1997. *Microenterprise Promotion Programs in the Developing Countries and their Relevance for the Hills of Nepal*. Paper presented at the East-West Center Participants Conference. Honolulu, Hawaii.

Leach, Edmund. 1976. *Culture and Communication. The Logic by Which Symbols are Connected*. Cambridge: Cambridge University Press.

Le Cup, Isabel. 1994. The role of marketing non-timber forest products in community development projects: Ayurvedic medicinal plants in Nepal. In: *Marketing of Multipurpose Tree Products in Asia*. Raintree, J.B. and H. A. Francisco. Pp. 269-275. Proceedings of an International Workshop held in Baguio City, Philippines, 6-9 December 1993. Winrock International, Bangkok.

Leslie, Joanne, Pelto, Gretel H. and Kathleen M. Rasmussen. 1988. Nutrition of women in developing countries. *Food and Nutrition Bulletin* 10(3):4-7.

Lett, James. 1990. Emics and etics: Notes on the epistemology of anthropology. In: *Emics and Etics. The Insider/Outsider Debate*. Thomas N. Headland, Kenneth L. Pike and Marvin Harris, eds. Pp. 127-142. Newbury Park: Sage Publications.

Levinson, David and Melvin Ember. 1996. *Encyclopedia of Cultural Anthropology. Volume 2*. New York: Henry Holt and Company.

Levitt, Marta J. 1988. *From Sickles to Scissors: Birth, Traditional Birth Attendants and Perinatal Health Development in Rural Nepal*. Ph.D. dissertation, University of Hawaii.

Lindemann, Shirley. 1977. The 'last course': nutrition and anthropology in Asia. In: *Nutrition and Anthropology in Action*. Thomas K. Fitzgerald, ed. Assen, Amsterdam: Van Gorcum.

Lintu, Leo. 1995a. Marketing non-wood forest products in developing countries. *UNASYLVA* 46(183):37-41.

_____. 1995b Trade and Marketing of Non-Wood Forest Products. FAO/Govt. of Indonesia Expert Consultation on Non-Wood Forest Products. Yogyakarta, 17-27 January 1995.

Mabberley, D. J. 1993. *The Plant-Book. A Portable Dictionary of the Higher Plants*. Cambridge: Cambridge University Press.

MacDonald, A.W. 1976. Preliminary notes on some *jhākri* of the Muglan. In: *Spirit Possession in the Nepal Himalayas*. John T. Hitchcock and Rex L. Jones, eds. Pp. 309-341. New Delhi: Vikas Publishing House Pvt. Ltd.

Majupuria, Trilok Chandra and D.P. Joshi. 1989. *Religious & Useful Plants of Nepal & India*. Bangkok: Craftsman Press.

Malhotra, K.C.; Deb, Debali; Dutta, M; Vasulu, T.S.; Yadav, G and M. Adhikari. 1993. The Role of Non-Timber Forest Products in Village Economies of Southwest Bengal. *ODI Rural Development Forestry Network Paper* 15d.1-8.

Malinowski, B. 1926. *Crime and Custom in Savage Society*. London: Kegan Paul, Trench and Trubner.

Malla, S.B. and P.R. Shakya. 1984-85. Medicinal plants. In: *Nepal-Nature's Paradise*. Majpurea, ed. Pp. 261-297. Bangkok: White Lotus Co. Ltd.

Malla, Y.B. 1992. *The Changing Role of the Forest Resource in the Hills of Nepal.* Ph.D. dissertation. Australian National University, Australia.

_____. 1982. *Medicinal plants of Nepal.* FAO. No. RAPA 64. Bangkok.

Malville, Nancy J. 1987. *Iron-Deficiency Anemia Among Village Women of the Middle Hills of Nepal.* Ph.D. dissertation. University of Colorado.

Manandhar, N.P. 1990a. Medico Botany of Gorkha District, Nepal- An elucidation of medicinal plants. *Institute Journal of Crude Drug Research* 28(1)17-25.

_____. 1990b. Traditional phyto therapy of Danuwar tribes of Kamlakhonj in Sindhuli District, Nepal. FITOTERAPIA LXI(4):325-331.

_____. 1990c. Folk-lore medicine of Chitwan District, Nepal. *Ethnobotany* 2:31-38.

_____. 1989a. Medicinal plants used by Chepang tribes of Makawanpur District, Nepal. *FITOTERAPIA* 15(1):61-68.

_____. 1989b. *Useful Wild Plants of Nepal.* Nepal Research Centre Publications No. 14. Stuttgart: Franz Steiner Verlag Wiesbaden GMBH.

_____. 1987. An ethnobotanical profile of Manang Valley, Nepal. *Journal of Economic Taxonomy Botany* 10(1):207-213.

_____. 1986. Ethnobotany of Jumla District, Nepal. *International Journal of Crude Drug Research* 24(2):81-89.

_____. 1985. Ethnobotanical notes on certain medicinal plants used by Tharus of Dang-Deokhuri District, Nepal. *International Journal of Crude Drug Research 23(4):153-159.*

_____. 1980. *Medicinal Plants of Nepali Himalaya.* Kathmandu: Ratna Pustak Bhandar.

Manderson, Lenore. 1981. Traditional food beliefs and critical life events in peninsular Malaysia. *Social Science Information* 20(6):947-975.

Mansberger, Joe R. 1991. *Ban Yatra: A Bio-Cultural Survey of Sacred Forests in Kathmandu Valley,* Ph.D. dissertation, University of Hawaii.

March, Kathryn S. 1987. Hospitality, women, and the efficacy of beer. *Food and Foodways* 1:352-387.

Martens, J. 1983. Forests and their destruction in the Himalayas of Nepal. *Nepal Research Center Miscellaneous Papers No. 35.* Kathmandu.

Maskarinec, Gregory. G. 1995. *The Rulings of the Night. An Ethnography of Nepalese Shaman Oral Texts.* Madison: The University of Wisconsin Press.

_____. 1993. Personal Communication. Honolulu, Hawaii.

McCay, Bonnie M. and James M. Acheson. 1987. *The Question of the Commons. The Culture and Ecology of Communal Resources.* Tucson: The University of Arizonia Press.

McDougal, Charles. 1979. *The Kulunge Rai: A Study in Kinship and Marriage Exchange.* Kathmandu: Ratna Pustak Bhandar.

McElroy, A. and P.K. Townsend. 1985. *Medical Anthropology in Ecological Perspective.* Boulder and London: Westview Press.

McGuire, Judith. 1993. Addressing micronutrient malnutrition. *SCN News No. 9.* Pp. 1-10. United Nations.

McGuire, Judith and Barry M. Popkin. 1990. Beating the zero-sum game: women and nutrition in the third world. In: *Women and Nutrition. ACC/SCN Symposium Report Nutrition Policy Discussion Paper No. 6.* Pp. 11-65. United Nations.

McNeely, J. A., Cronin, E. W. and H. B. Emery. 1973. The Yeti- not a snowman. *ORYX* 12:65-73.

McNeely, Jeffrey A. 1995. *Expanding Partnerships in Conservation.* Washington D.C.: Island Press.

Mead, Margaret. 1945. *Manual for the Study of Food Habits. Bulletin 111.* Washington, D.C.: Committee on Food Habits, National Research Council.

Meilleur, Brien A. 1994. In search of "keystone societies." In: *Eating on the Wild Side. The harmacologic, Ecologic, and Social Implications of Using Noncultigens*. Nina L. Etkin, ed. Pp. 259-279. Tucson: The University of Arizonia Press.

Messer, Ellen. 1989. The relevance of time-allocation analyses for nutritional anthropology. The relationship of time and household organization to nutrient intake and status. In: *Research Methods in Nutritional Anthropology*. Gretel H. Pelto, Pertti J. Pelto and Ellen Messer, eds. Pp. 82-125. The United Nations University.

Messer, Ellen and Harriet Kuhnlein. 1986. Traditional foods. In: *Training Manual in Nutritional Anthropology*. Sara A. Quandt and Cheryl Ritenbaugh, eds. Pp. 66-81. A special publication of the AAA number 20.

Messerschmidt, Donald, A. 1995. Local Traditions and Community Forestry Management: A View from Nepal. In: The Cultural Dimension of Developing Indigenous Knowledge Systems. D. Michael Warren, L. Jan Slikkerveer and David Brokensha, eds. Pp. 231-244. London: LTD.

_____. 1987. Conservation and society in Nepal: Traditional forest management and innovative development. In: *Lands at Risk in the Third World. Local-Level Perspectives*. Little, Peter D., Horowitz, Michael A. and A. Endre Nyerges, eds. Pp. 373-397. Boulder and London: Westview Press.

_____. 1986. People and Resources in Nepal: Customary Resource Management Systems of the Upper Kali Gandaki. In: Proceedings of the Conference on Common Property Management. Pp. 455-480 (see Arnold and Campbell for full citation).

Metz, John J. 1990. Conservation practices at an upper-elevation village of West Nepal. *Mountain Research and Development* 10(1):7-15.

_____. 1989. *The Goth System of Resource Use at Chimkhola, Nepal*. Ph.D. dissertation, Madison: University of Wisconsin.

Milton, Kay. 1996. *Environmentalism and Cultural Theory. Exploring the role of anthropology in environmental discourse*. London: Routledge.

Miracle, M.P. 1961. Seasonal hunger: a vague concept and an unexplored problem. *Bulletin de l'IIFAN* XXIII, Series B: 271-83.

Moench, Marcus and J. Bandyopadhyay. 1986. People-forest interaction: a neglected parameter in Himalayan forest management. *Mountain Research and Development* 6(1):3-16.

Molnar, Augusta. 1981a. Economic strategies and ecological constraints: case of the Kham Magar of Northwest Nepal. In: *Asian Highland Societies in Anthropological Perspective*. Christoph von Furer-Haimendorf, ed. Pp. 20-51. New Delhi: Sterling Publishers.

_____. 1981b. Nepal. *The Dynamics of Traditional Systems of Forest Management: Implications for the Community Forestry Development and Training Project*. World Bank Report.

Mookerji, Radha Kumud. 1956. *Ancient India*. Allahabad: Indian Press Publishing Private Limited.

Moran, Emilio F. 1990. Ecosystem ecology in biology and anthropology: a critical assessment. In: *The Ecosystem Approach in Anthropology. From Concept to Practice*. Emilio F. Moran, ed. Pp. 3-40. Ann Arbor: The University of Michigan Press.

Morris, Brian. 1982. Forest traders. A socio-economic study of the hill Pandaram. *London School of Economics Monograph on Social Anthropology No. 55*. New Jersey: Athlone Press

Murai, Mary, Pen, Florence and Carey D. Miller. 1958. *Some Tropical South Pacific Foods*. Honolulu: University of Hawaii Press.

Nabarro, D., Cassels, C., Pant, M. and Wijga, A. 1987. The Impact of Integrated Rural Development: The Kosi Hill Area Rural Development Programme, East Nepal. Liverpool: School of Tropical Medicine.

Nabarro, David, Cassels, Claudia and Maheṣh Pant. 1989. *Coping Strategies of Households in the Hills of Nepal: Can Development Initiatives Help? IDS Bulletin* 20(2):68-74. Institute Development Studies, Sussex.

Nabarro, David. 1984. Social, economic, health and environmental determinants of nutritional status. *Food and Nutrition Bulletin* 6(1):18-32.

_____. 1982 Influences on child nutrition in Eastern Nepal. *Journal of the Institute of Medicine* 4(1):47-66.

Nag, M.J., White, B. and R.C. Peet. 1978. An anthropological approach to the study of the economic value of children in Java and Nepal. *Current Anthropology* 19:234-306.

NAS. 1980. *Proceedings: International Workshop on Energy Survey Methodologies for Developing Countries*. National Academy of Sciences. Washington, D.C.: National Academy Press.

Natadecha, Poranee. 1988. Buddhist religion and scientific ecology as convergent perceptions of nature. In: *Essays on Perceiving Nature*. Diane M. Deluca, ed. Pp. 113-118. Honolulu: Perceiving Nature Conference Committee.

National Vitamin A Workshop. 1992. *Report on National Vitamin A Workshop held 11/12 February 1992*. Organized by the Ministry of Health. His Majesty's Government of Nepal. Sponsored by USAID/VITAL-UNICEF-VACSP/NNIPS.

Nazarea-Sandoval, Virginia D. 1995. *Local Knowledge and Agricultural Decision Making in the Philippines. Class, Gender, and Resistance*. Ithaca: Cornell University Press.

Negi, S.S. 1992. *Minor Forest Products of India*. Delhi: Periodical Experts Book Agency.

Nepali, Rohit K. 1993. Culturally appropriate development: An anthropological study of villages in Eastern Nepal. In: *Anthropology of Tibet and the Himalaya*. Charles Ramble and Martin Brauen, eds. Pp. 238-247. Zurich: Ethnological Museum of the University of Zurich.

Neri, Bayani S. 1994. Philippines. In: *Non-Wood Forest Products in Asia*. Patrick B. Durst, Ward Ulrich and M. Kashio, eds. Pp. 97-116. RAPA Publications 1994/28 FAO: Bangkok.

NERRA/IIRR. 1992. Regenerative Agriculture Technologies for the Hill Farmers of Nepal. Kathmandu.

Nietschmann, Bernard. 1973. *Between Land and Water*. New York: Seminar Press.

NNCC. 1978 National Nutrition Co-ordination Committee, National Nutrition Strategies, Pokhara, Kathmandu.

NPC. 1991. *Approach to the Eighth Plan (1992-97)*. National Planning Commission. Kathmandu: His Majesty's Government.

NTFPs. 1994. *National Seminar on Non-Timber Forest Products: Medicinal and Aromatic Plants. Proceedings*. Kathmandu. September 11-12, 1994. Ministry of Forests and Soil Conservation and Herbs Production and Processing Co. LTD.

NWFPs. 1995a. *Beyond Timber: Social, Economic and Cultural Dimensions of Non-Wood Forest Products in Asia and the Pacific*. Patrick B. Durst and Ann Bishop, eds. Bangkok: FAO

NWFPs. 1995b. *Report of the International Expert Consultation on Non-Wood Forest Products*. Yogyakarta, Indonesia. 17-27 January 1995. FAO: Rome.

Ogden, C. 1990. Building nutritional considerations into forestry development efforts. *Unasylva* 41:20-28.

Oppitz, Michael. 1973. Myths and facts: reconsidering some data concerning the clan history of the Sherpa. In: *Contribution to the Anthropology of Nepal*. Christoph von Furer Haimendorf, ed. Pp. 232-243. Warminister: Aris and Phillips, Ltd.

_____. 1968. *Geschichte und Sozialordnung der Sherpa*. Innsbruck and Munich: Universitats Verlag Wagner.

Orlove, Benjamin S. 1980. Ecological anthropology. *Annual Review of Anthropology* 9:235-273.

Ortiz, Sutti. 1990. Uncertainty reducing strategies and unsteady states: Labor contracts in coffee agriculture. In: *Risk and Uncertainty in Tribal and Peasant Economies.* Elizabeth Cashdan, ed. Pp. 303-317. Boulder, Colorado: Westview Press.

Ortner, Sherry B. 1995. The case of the disappearing shamans, or no individualism, no relationalism. *ETHOS* 23(3):355-390.

_____. 1984. Theory in anthropology since the Sixties. *Comparative Studies in Society and History.* 26(1):126-165.

_____. 1978. *Sherpas Through Their Rituals.* Cambridge: Cambridge University Press.

_____. 1974. Is female to male as nature is to culture? In: *Woman, Culture and Society.* M. Rosaldo and L. Lamphere, eds. Stanford, CA: Stanford University Press.

_____. 1973. Sherpa purity. *American Anthropologist* 75(i):49-63.

Otto, J. S. and N. E. Anderson. 1982. Slash-and-burn cultivation in the highlands south: a problem in comparative agricultural history. *Society for Comparative Study of Society and History* volume 24:131-147.

Pandey, Mrigendra R. 1984. Domestic smoke pollution and chronic bronchitis in a rural community in the hill region of Nepal. *Thorax* 39:337-339.

Pandey, Shanta. 1993. Women, environment and local initiatives: Factors affecting the degree of successful management of forest resources. *Himalayan Research Bulletin 13(1-2): 54-59.*

_____. 1991. Criteria and indicators for monitoring and evaluation of the social and administrative aspects of improved cookstove programs (ICPs). *Background Report No. 3.* Environment and Policy Institute, East-West Center, Honolulu.

_____. 1987. *A Methodology To Study Women In Resource Management and a Case Study of Women in Forest Management in Dhading District, Nepal.* International Center for Integrated Mountain Development, Nepal.

Panter-Brick, C. 1989. Motherhood and subsistence work: the Tamang of rural Nepal. *Journal of Human Ecology* 17(2):205-228.

Parkin, Robert. 1992. *The Munda of Central India. An Account of their Social Organization.* Delhi: Oxford University Press.

Paul, Robert A. 1982. *The Tibetan Symbolic World. Psychoanalytic Explorations.* Chicago: The University of Chicago Press.

_____. 1976. Some observations on Sherpa shamanism. In: *Spirit Possession in the Nepal Himalayas.* John T. Hitchcock & Rex L. Jones, eds. Pp. 141-151. New Delhi: Vikas Publishing House Pvt LTD.

Pei, Sheng-ji. 1985. Some effects of the Dai people's cultural beliefs and practices upon the plant environment of Xishuangbanna, Yunnan Province, Southwest Asia. In: *Cultural Values and Human Ecology in Southeast Asia.* Karl L. Hutterer, A. Terry Rambo and George Lovelace, eds. Pp. 321-339. Ann Arbor: University of Michigan Center for South and Southeast Asia Studies, Paper No. 27.

Pelto, Gretel H. 1984. Ethnographic studies of the effects of food availability and infant feeding practices. *Food and Nutrition Bulletin* 6(1):33-43.

Pelto, Gretel H. and Norge W. Jerome. 1978. Intracultural Diversity and Nutritional Anthropology. In: *Health and the Human Condition. Perspective on Medical Anthropology.* Michael H. Logan and Edward E. Hunt, Jr. eds. Pp. 322-328. Duxbury Press: Massachusetts.

Peluso, Nancy L. 1992. The political ecology of extraction and extractive reserves in East Kalimantan, Indonesia. *Development and Change* 23(4):49-74.

_____. 1983 Markets and Merchants: *The Forest Products Trade of East Kalimantan in Historical Perspective.* MS thesis, Cornell University.

Piazza, Alan. 1986. *Food Consumption and Nutritional Status in the PRC*. Boulder: Westview Press.

Pigg, Stacy L. 1989. Here, there and everywhere: place and person in Nepalese explanations of illness. *Himalayan Research Bulletin* 9(2):16-24.

Piller, Andrew B. 1986. *Food Intake and Energy Output in Eastern Nepal*, MS thesis, University of Hawaii at Manoa.

Pimentel, David, McNair, Michael, Buck, Louise, Pimentel, Marcia and Jeremy Kamil. 1997. The value of forests to world food security. *Human Ecology* 25(1):91-120.

Pinkerton, Evelyn. 1987. Intercepting the state: Dramatic processes in the assertion of local co-management rights. In: *The Question of the Commons. The Culture and Ecology of Communal Resources*. Bonnie J. McCay and James M. Acheson, eds. Pp. 344-369. Tucson: The University of Arizonia Press.

Pitt, David C. 1986. Crisis, pseudocrisis or supercrisis. poverty, women and young people in the Himalaya. a survey or recent developments. *Mountain Research and Development* 6(2):119-131.

Plattner, Stuart. 1989. *Economic Anthropology*. Stanford: Stanford University Press.

Plotkin, Mark and Lisa Famolare, eds. 1992. *Sustainable Harvest and Marketing of Rain Forest Products*. Washington, D.C.: Island Press.

Plotkin, Mark. 1988. The outlook for new agricultural and industrial products from the tropics. In: *Biodiversity*. E.O. Wilson, ed. Pp. 106-116. Washington, D.C.: National Academy Press.

Pohle, Perdita. 1990. Useful plants of the Manang District. A contribution to the Ethnobotany of the Nepal Himalaya. Nepal Research Centre Publications No. 16. Stuttgart: Franz Steiner Verlag Wiesbaden GMBH.

Pollock, Nancy J. 1992. *These Roots Remain: Food Habits in Islands of the Central and Eastern Pacific Since Western Contact*. Laie, HI: Institute for Polynesian Studies.

Prasai, Yogendra, Gronow, Jane, Bhuju, Ukesh R. and Sharmila Prasai. 1987. Women's Participation on Forest Committees. A Case Study. *Forestry Research Paper Series Number 11*. HMG-USAID-GTZ-IDRC-Ford-Winrock Project.

Quanine, Jannat. 1990. Women and nutrition: Grameen Bank Experience. In: *Women and Nutrition*. Pp. 109-117. ACC/SCN Symposium Report. Nutrition Policy Discussion Paper No. 6. United Nations.

Raintree, J.B.(ed). 1986. *An Introduction to Agroforestry Diagnosis and Design*. Nairobi: International Council on Agroforestry Research.

Raintree, J.B. and Francisco, H.A. 1994. *Marketing of Multipurpose Tree Products in Asia*. Proceedings of an International Workshop held in Baguio City, Philippines, 6-9 December 1993. Winrock International, Bangkok.

Rajbhandari, Keshab R. 1991. *Grassland Ecology and Preliminary Studies of Bamboo in the Apsuwa Valley*. The Makalu-Barun Conservation Project Working Paper Publications Series Report 13. Kathmandu.

Rajbhandary, T.K. and J.M. Bajracharya. 1994. *National Status on NTFPs: Medicinal and aromatic plants*. National Seminar on Non-Timber Forest Products: Medicinal and Aromatic Plants. Proceedings. Pp. 8-15. September 11-12, 1994. Kathmandu.

Ramble, Charles. 1993. Interview. Reflections of a Plant-Hunter in Nepal: An interview with Dr. Tirtha Bahadur Shrestha. *European Bulletin of Himalayan Research* 5:34-40.

Ramble, Charles and Chandi P. Chapagain. 1990. Preliminary Notes on the Cultural Dimension of Conservation. Makalu Barun Conservation Project *Working Paper Publication Series*.

Rambo, A. Terry. 1980. Fire and the Energy Efficiency of Swidden Agriculture. *East-West Environment and Policy Institute Reprint No. 55*. East-West Center, Honolulu.

Rappaport, Roy A. 1979. *Ecology, Meaning and Religion*. Richmond, California: North Atlantic Books.

_____. 1971. Nature, Culture and Ecological Anthropology. *Man, Culture and Society*. Harry L. Shapiro, Ed. London: Oxford University Press.

_____. 1968 *Pigs for the Ancestors*. New Haven: Yale University Press.

Rawal, Rana B. 1994. Commercialization of aromatic and medicinal plants in Nepal. In: *Beyond Timber: Social, Economic and Cultural Dimensions of Non-Wood Forest Products in Asia and the Pacific*. Patrick B. Durst and Ann Bishop, eds. Pp. 149-55. Bangkok: FAO

Regmi, Mahesh C. 1978. *Land Tenure and Taxation in Nepal*. Kathmandu: Ratna Pustak Bhandar.

Reichel-Dolmatoff, G. 1976. Cosmology as ecological analysis: a view from the rain forest. *Man, New Series* 2(3):307-318.

Reid, Holly F., Smith, Kirk R. and Bageshwari Sherchand. 1986. Indoor smoke exposures from traditional and improved cookstoves: Comparisons among rural Nepali women. *East-West Environment and Policy Institute Reprint No. 98*.

Reinhard, Johan. 1987. The sacred Himalaya. *The American Alpine Journal* 29(61):123-132.

_____. 1978. Khembalung: the hidden valley. *Kailash* 6(1):5-35.

Richard, Audrey I. 1939. *Land, Labour and Diet in Northern Rhodesia*. London: Oxford University Press.

Richards, Paul. 1980. Community environmental knowledge in African rural development. In: *Indigenous Knowledge Systems and Development*. David W. Brokensha, D.M. Warren and Oswald Werner, eds. Pp. 183-196. Lanham: University Press of America.

Rindos, David. 1984. *The Origins of Agriculture. An Evolutionary Perspective*. Orlando: Academic Press, Inc.

Rocheleau, D.E., Wachira, D.K., Malaret, L. and B.M. Wanjohi. 1989. Local Knowledge for Agroforestry and Native Plants, pp: 14-23, In: *Farmer First: Farmer Innovations and Agricultural Research*. A. Pacey and LA Thrupp, eds. London: IT.

Rock, Joseph F. 1930. Seeking the mountains of mystery. An expedition on the China-Tibet frontier to the unexplored Amnyi Machen range, one of whose peaks rivals Everest. *The National Geographic Magazine* 57(2):

Rodda, Annabel. 1991. *Women and the Environment*. London: Zed Books Ltd.

Rosenberg, E.M. 1980. Demographic effects of sex-differential nutrition. In: *Contemporary Approaches to Diet & Culture*. Norge W. Jerome, Randy F. Kandel and Gretel H. Pelto, eds. Pp. 181-203. Pleasantville, NY: Redgrave Publishing Company.

Russell, Andrew. 1992. The hills are alive with the sense of movement: migration and identity amongst the Yakha of East Nepal. *Himalayan Research Bulletin* 12(1-2):35-43.

Russell, W.M.S. 1968. The slash-and-burn technique. *Natural History*. Pp. 86-101.

Ryder, R. 1974. *Speciesism: The Ethics of Vivisection*. Edinburgh: Scottish Society for the Prevention of Vivisection.

Sacherer, Janice. 1979a. The high-altitude ethnobotany of the Rolwaling Sherpas. *Contributions to Nepalese Studies* 6(2):45-64.

_____. 1979b. *Practical Problems in Development in Two Panchāyats in North Central Nepal: A Baseline Study*. Kathmandu: SATA.

Sagant, Philippe. 1976. Becoming a Limbu priest: ethnographic notes. In: *Spirit Possession in the Nepal Himalayas*. John T. Hitchcock & Rex L. Jones, eds. Pp. 56-99. New Delhi: Vikas Publishing House Pvt LTD.

Samuel, Geoffrey. 1993. *Civilized Shamans. Buddhism in Tibetan Societies*. Washington: Smithsonian Institution Press.

Sanwal, Mukul. 1989. What we know about mountain development: common property, investment priorities, and institutional arrangements. *Mountain Research and Development* 9(1):3-14.

Sauer, Carl O. 1969. *Agricultural Origins and Dispersals. The Domestication of Animals and Foodstuffs.* Second Edition. Cambridge: The M.I.T. Press.

Saul, Rebecca. 1994. Indigenous forest knowledge: Factors influencing its social distribution. In: *The Anthropology of Nepal, Peoples, Problems and Processes.* Michael Allen, ed. Pp. 136-144. Kathmandu: Mandala Book Point.

Savada, A.M. 1993. *Nepal and Bhutan: Country Studies. Area Handbook Series.* Washington D.C.: U.S. Government Printing Office.

Schelhas, John and Russell Greenberg. 1996. Introduction: the value of forest patches. In: *Forest Patches in Tropical Landscapes.* John Schelhas and Russell Greenberg, eds. Pp. xv-xxxvi. Washington, DC: Island Press.

Schmidt-Vogt. D. 1999. Swidden farming and fallow vegetation in Northern Thailand. *Geoecological Research Vol. 8.* Franz Steiner Verlag. Stuttgart.

Schuler, Sidney R. 1981. The women of Baragaon. In: *The Status of Women in Nepal.* Volume II. Part 2. CEDA. Tribuhuvan University, Kathmandu.

Schulthess, W. 1967. Yak and *chauri* in Nepal. *World Review Animal Production* 3:88.

Scoones, I. Melnyk, M. and J.N. Pretty. 1992. *The Hidden Harvest: Wild Foods and Agricultural Systems, a Literature Review and Annotated Bibliography.* London: International Institute for Environment and Development (IIED).

Seeland, Klaus. 1997. Sociological Observations on "Community Forestry" in Nepal. In: *Perspectives on History in Karakorum, Hindukush and Himalaya.* J. Stellrecht, ed. Kolu: R. Koppe Verlag.

_____. 1993. Sanskritization and environmental perception among Tibeto-Burman speaking groups. In: *Anthropology of Tibet and the Himalaya.* Charles Ramble and Martin Brauen, eds. Pp. 354-361. Zurich: Ethnological Museum of the University of Zurich.

_____. 1985. The use of bamboo in a Rai village. *Journal of the Nepal Research Centre* 4:175-187.

Seeley, Janet. 1989. *Women and Extension Services in Lumle Agricultural Centre. Research and Extension Command Areas.* Technical Paper 16. Lumle Agricultural Centre.

Sharma, Pitamber. 1989. Urbanization in Nepal. *Papers of The East-West Population Institute* Number 110, East-West Center, Honolulu.

Sharma, Prayag R., Dahal, Dilli R. and J. M. Gurung. 1991. *Cultural Systems and Resources.* King Mahendra Trust for Nature Conservation. Arun III: Management of Basinwide Environmental Impacts Study.

Sharma, Y.M.L. 1982. Some aspects of bamboo: In: *Asia and the Pacific. No. RAPA 57.* Bangkok: FAO.

Shepherd, K.R. 1985 Management of the Forest and Agricultural Systems Together for Stabilization and Productivity in the Middle Hills of Nepal. Paper Presented at the International Workshop on Water-shed Management in the Hindu Kush Himalayan Region, ICIMOD/CISNAR, Chengdu, China.

Shils, Maurice and Vernon R. Young. 1988. *Modern Nutrition in Health and Disease.* Seventh Edition. Philadelphia: Lea & Febiger.

Shiva, Vandana. 1989. *Staying Alive. Women, Ecology and Development.* London: Zed Books.

Shrestha, T.B., Sherpa, L.N., Banskota, K. and R.K. Nepali. 1990. *The Makalu-Barun National Park and Conservation Area Management Plan.* Department of National Parks, Wildlife Conservation, HMG/Nepal and Woodlands Mountain Institute, Kathmandu.

Shrestha, Tirtha Bahadur. 1989. Development Ecology of the Arun River Basin in Nepal. *International Centre For Integrated Mountain Development,* Kathmandu.

Simoons, Frederick J. 1994. *Eat Not This Flesh. Food Avoidances from Prehistory to the Present.* Madison: The University of Wisconsin Press.

_____. 1991 *Food in China. A Cultural and Historical Inquiry.* Boca Raton: CRC Press.

Singh, Subarna L. 1989. Overview of the girl child in Nepal. The National Seminar on the 1990 SAARC Year of the Girl Child held 25-27 September 1989, in Kathmandu, Nepal. WSCC/UNICEF.

Smith, Kirk. 1997. Indoor air quality and child health in developing countries. Letter. University of California, Berkeley.

Sponsel, Leslie E. forthcoming. Religion and environmental ethics: spiritual ecology. In: *Ecocide or Ecosanity? An Ecological Anthropology of Diversity.* Book manuscript.

_____. 1997. Ecological anthropology. In: *Dictionary of Cultural Anthropology.* Thomas Barfield, ed. Blackwell Publishers (in press).

_____. 1992. The environmental history of Amazonia: Natural and human disturbances and the ecological transition. In: *Changing Tropical Forests. Historical Perspectives on Today's Challenges in Central and South America.* Harold K. Steen and Richard P. Tucker, eds. Pp. 233-251. Sponsored by the Forest History Society and IUFRO Forest History group.

_____. 1987. Cultural ecology and environmental education. *Journal of Environmental Education* 19(1):31-42.

Sponsel, Leslie E. and Poranee Natadecha-Sponsel. 1993. The potential contribution of Buddhism in developing an environmental ethic for the conservation of biodiversity. *Ethics, Religion and Biodiversity. Relations Between Conservation and Cultural Values.* Lawrence S. Hamilton, ed. Pp. 75-97. Cambridge: The White Horse Press.

Srinivas, M.N. 1956. A note on Sanskritization and Westernization. *Far Eastern Quarterly* 15:481-482.

Standal, Bluebell. 1997. Personal communication during meetings in July. Honolulu, Hawaii.

Stevens, Stanley F. 1997. Consultation, Co-Management and Conflict in Sagarmatha (Mount Everest) National Park, Nepal. In: *Conservation Through Cultural Survival. Indigenous Peoples and Protected Areas.* Stanley Stevens, ed. Pp. 63-97. Washington, D.C: Island Press.

_____. 1993. *Claiming the High Ground. Sherpas, Subsistence and Environmental Change in the Highest Himalaya.* Berkeley: University of California Press.

Stevens, Stanley F. and Mingma N. Sherpa. 1993. Indigenous peoples and protected areas: New approaches to conservation in highland Nepal. *Parks, Peaks and People.* Lawrence S. Hamilton, Daniel P. Bauer, and Helen F. Takeuchi, eds. Pp. 73-88. Honolulu: East-West Center.

Steward, Julian. 1955. *The Theory of Culture Change.* Urbana: University of Illinois Press.

Stone, Linda. 1983. Hierarchy and food in Nepalese healing rituals. *Social Science and Medicine* 17(14):971-978.

Strickland, S.S. 1990. Traditional economies and patterns of nutritional disease. In: *Diet and Disease.* Harrison, G.A. and J.C. Waterlow, eds. Pp. 209-239. Cambridge: Cambridge University Press.

Subba, K.J. 1996. *Non-Wood Forest Products of Bhutan.* RAP Publication: FAO.

Thomas-Slayter, Barbara and Nina Bhatt. 1994. Land, livestock, and livelihoods: changing dynamics of gender, caste and ethnicity in a Nepalese village. *Human Ecology* 22(4):467-494.

Thompson, M., Warburton, M. and T. Hatley. 1986. *Uncertainty on a Himalayan Scale.* London: Ethnographica.

Toba, Sueyoshi. 1975. Plant names in Khaling. A study in ethnobotany and village economy. *KAILASH* 3(2):145-169.

Tucker, Mary E. 1997. The emerging alliance of religion and ecology. *Worldviews. Environment, Culture, Religion* 1(1):3-24.

Turner, Ralph L. 1980[1931]. *A Comparative and Etymological Dictionary of the Nepali Language*. New Delhi: Allied Publishers Private Limited.

UMN. 1987. *Health for the Hills. An Evaluation of the UMN's Okhaldunga CHP, 1978-85 and Current Baseline Data on the Project Area*. Studies and Evaluation Programme Rural Development Centre. United Mission to Nepal, Kathmandu.

UNDP. 1987. Cultivation of Medicinal and Aromatic Plants. United Nations Development Project/Food and Agricultural Organization, Rome.

UNESCO 1991. *Statistical Yearbook for Asia and the Pacific*. United Nations Economic and Social Commission for Asia and the Pacific. Bangkok.

UNFPA. 1992. National Fertility Family Planning and Health Survey.

UNICEF. 1987. *Children and Women Of Nepal. A Situation Analysis*. United Nations Children's Fund.

Vansittart, E. 1915. *Gurkhas*. Calcutta: Superintendent of Government Printing.

Vayda, A.P. and B.J. McCay. 1977. Problems in the identification of Environmental Problems. In: *Subsistence and Survival: Rural Ecology in the Pacific*. T.P. Bayliss Smith and R.G. Feachem, eds. London: Academic Press.

Velarde, Nancy. 1990. Putting household food security aspects into farming systems research. *Forest, Trees and People Newsletter* 11:12-16.

Vincent, Joan. 1986. System and process 1974-1985. *Annual Review of Anthropology* 15:99-119.

Wahlquist, Hakan. 1981. Rural indebtedness in North-Eastern Nepal. Part I: The ethnography and its implications. *ETHNOS* 46:207-238.

Wallace, Michael B. 1987. *Community Forestry in Nepal: Too Little, Too Late?* Research Report Series. HMG-USAID-GTZ-IDRC-FORD-Winrock Project. Kathmandu.

Wangmo, Sonam. 1990. The Brokpas: A semi-nomadic people in Eastern Bhutan. In: *Himalayan Environment and Culture*. N.K. Rustomji and Charles Ramble, eds. Pp. 141-158. Shimla: Indian Institute of Advanced Study.

Wealth of India. 1985. *A Dictionary of Indian Raw Materials and Industrial Products. Raw Materials*. Volume I:A. Revised. New Delhi: Publications and Information Directorate, CSIR.

Weber, Will. 1991. Enduring peaks and changing cultures. The Sherpas and Sagarmatha (Mount Everest) national park. In: *Resident Peoples and National Parks. Social Dilemmas and Strategies in International Conservation*. Patrick C. West and Steven R. Brechin, eds. Pp. 206-214. Tucson: The University of Arizionia Press.

Wehmeyer, A.S. 1966. The nutrient composition of some edible wild fruits found in the Transvaal. *Southern African Medical Journal*.

Weitz, Charles A., Pawson, Ivan G., Weitz, M. Velma, Lang, Selwyn D.R. and Ann Lang. 1978. Cultural factors affecting the demographic structure of a high-altitude Nepalese population. *Social Biology* 25(3):179-195.

Wheeler, E.F. and M. Abdullah 1988. Food allocation within the family: response to fluctuating food supply and food needs. In: *Coping With Uncertainty in Food Supply*. I. de Garine and G.A. Harrison, eds. Pp. 437-451. Oxford: Clarendon Press.

White, Leslie A. 1959. *The Evolution of Culture*. New York: McGraw-Hill.

Widdowson, Elsie M. 1991. Contemporary Human Diets and their Relation to Health and Growth: Overview and Conclusions. *Philosophical Transactions Royal Society of London. Biological Sciences* 334:289-295.

Wild Edible Plants of Nepal. 1982. *Wild Edible Plants of Nepal*. Bulletin No. 9. Department of Medicinal Plants, His Majesty's Government, Nepal.

Wilk. Richard. R. 1996. *Economies and Cultures. Foundations of Economic Anthropology.* Boulder, Colorado: Westview Press.

Wilson, Christine S. 1974. Child following: A technique for learning food and nutrient intakes. *Environmental Child Health* 20:9-14.

World Bank. 1979. *Nepal: Development Performance and Prospects. A World Bank Country Study, South Asia Regional Office*. Washington, D.C.: The World Bank.

Worth, Robert M. and Narayan K. Shah. 1969. *Nepal Health Survey 1965-1966*. Honolulu: University of Hawaii Press.

Wright, Gary A. and Jane D. Dirks. 1983. Myth as environmental message. *ETHNOS* 3-4:160-176.

Wu, David Yen-Ho. 1979. The Ever-Popular Yin-Yang Way to Health. *East-West Perspectives* (Winter Issue). Pp. 89-92.

Wyatt-Smith, John 1982. The agricultural system in the hills of Nepal: The ratio of agricultural to forest land and the problem of animal fodder. *Agricultural Projects Services Centre Occasional Paper 1*. Kathmandu.

Yonzon, Pralad B. and Malcolm L. Hunter, Jr. 1991. Cheese, tourists, and red pandas in the Nepal Himalayas. *Conservation Biology* 5(2):196-202.

Yonzon, Pralad. 1993. Raiders of the park. *HIMAL*:22-23.

Zangbu, Ngawang T. and Frances Klatzel. 1995. *Stories and Customs of the Sherpas*. Kathmandu: Mandala Book Point.

Zeitlin, Marian F. and Laurine V. Brown. 1992. Integrating diet quality and food safety into food security programmes. *Nutrition Consultants' Report Series 91*. Rome: FAO.

Zimmermann, F. 1987. *The Jungle and the Aroma of Meats: An Ecological Theme in Hindu Medicine*. Berkeley: University of California Press.

Zurick, David N. 1988. Resource needs and land stress in Rapti Zone, Nepal. *Professional Geographer* 40(4):428-444.

APPENDIX

'Wild' and 'Semi-Wild' Plants[1]

***Allium hypsistum*; Alliaceae; Aromatic Leaf Garlic; *Jimbu* (N), *Ramba* (S)**
A wild growing aromatic herb. **Uses:** 1) Dried leaves used as seasoning herb, spice and condiment. 2) Medicinal value.

***Amomum aromaticum* Roxb.; Zingiberaceae; *Alainchi* (N, R, S)**
A leafy perennial with a dark brown fruit. Located close to Tamkhu. Available August to December. **Uses:** 1) The white flesh of the fruit is eaten by both Rais and Sherpas. 2) Medicine (aromatic seeds) for colds and coughs, used only by Sherpas. 3) Traded by the Rais.

***Arisaema flavum*; Araceae; *Toho* (N, R); *Phi tho* (S)**
A yellowish-brown tuber plant, similar to taro (cobra plant) with dark and light green spotted leaves and a white stem. Grows to a height of one meter. Found at both the *lekh* and *jangal*, Dhap, Deorali, and Metalung (nearby Gongtala). June-September available. **Uses:** 1) Bear and wild boar food. 2) Used for *raksī* (Sherpas only) 3) Medicine for stomach ailments (Sherpas only). 4) Food: a. The root is cooked like potatoes and mashed. The Sherpas said, that it must be eaten with soup, or it can numb the mouth because it is a poisonous plant. b. The root is made into bread by the Rais. 5) Insecticide (clothing)

***Artemisia vulgaris* L.; Compositae (Wormwood); *Titepāti* (N); *Sibuma* (R); *Khenba* (S);**
An aromatic shrub with olive green leaves and a brown stem. The plant has a nice fragrance. Found at Yangden. March-November, August-September (flowers). **Uses:** 1) Religious plant, used as incense. 2) Yeast starter for *chang* (flower). 3) Sheep fodder.

---

1   The plants are listed in the following order: generic name, species name, family name, N- Nepali, R-Kulung Rai and S- Sherpa, based on the farmers in the research area. All of these plants were collected and identified in Kathmandu (See Sources of Identification).

Arundinaria aristatla; Gamble; Gramineae; Himalayan Bamboo; *Māliṅgo* (N); *Mikpo* (R); *Chokting* (S)

A tall shruby bamboo. Found above Dobatak, at Dhap and Merengma.

July-August available. **Uses:** 1) House building material (*citra*) 2) Tender spring shoot (edible), eaten only by Sherpas. 3) Matured culms are woven into baskets. 4) Fodder (leaves) cows and goats.

Arundinaria falcata; Gramineae; *Kālo nigālo* (N); *Dumdhu* (R); *Funyuk* (S)

The small bamboo which grows 2-4 m tall and is 1-1.5 cm. in diameter. Deorali (the *jaṅgal* north of Gongtala) July-September available. **Uses:** 1) Young shoots are eaten as food (*tusa*) 2) Woven into mats for the houses and animal shelters. 3) Cow, goat and sheep fodder.

Berginia purpurascens; Saxifragaceae; *Pakhanbed; Pakambet* (N, R, S)

An herb with large green leaves and a brown root. Found at Dorongbuk (*lekh*). **Uses:** 1) Medicine(root)- for severe diarrhea, it is drunk as a tea only by Sherpas. Has a bitter taste. 2) It is traded to India. 3) The leaves are put over the cooked maize or rice, and then ashes are placed on the top to keep the food warm. 4) Given to a woman after childbirth to give her energy. (Chopped into small pieces, then it is made into a powder, and cooked with butter, sugar may also be added); reported by a Sherpa from the Solu region. 5) Sheep fodder.

Bistorta spp.; Polygonaceae; *Rambu* (N, R, S)

Flowers are fuchia and the stem is purplish with green leaves. There is no fragrance. Grows to a height of 6-14 inches. Open space, grows by itself. Found at the *lekh* (about 5,500 m). June-September. **Uses:** 1) Flowers are eaten. 2) Made into flour which is similar to millet (flower). 3) Rais eat as a vegetable. 4) Animal fodder.

Bombay spp.; Singane (N); *Pakpo* (R); *Napshal* (S)

Location: Above Gongtala (Tokpu). August-September. **Uses:** 1) Fodder: cows, goats and sheep. 2) Building material for houses (*citra*). 3) Woven baskets (*doko*).

Brassaiopsis schefflera sp.; Araliaceae; *Budopo* (R); *Ramar* (S)

Olive-green leaves. Grows to a height of 10-12 m. Found at the *lekh*, Dhap, Dobrato. March-May. **Uses:** Cow fodder, "compared to all the fodders, this is considered the most nutritious."

Cardamine L. hirsuta; Cruciferae; Bittercress; *Mangasag* (N, R)

(**Mangacherman**(S)- *cherman* means vegetable in Sherpa). A plant with green leaves which are light purple on the edge and has purple and white flowers. Two types: 1) *Lekh* (5-6 inches), Kholā Kharka and Dorongbuk. 2) *Jaṅgal* (about 1 m). June to September. **Uses:** 1) The flowers and leaves are eaten as a vegetable by the

Rais. At the *lekh*, it is first dried in the sun to remove the bitter taste. 2) Sheep and goat fodder (leaves).

Castanopsis hystryx; **Fagaceae;** *Patle*(**N**); *Dokshim* (**S**); *Sevla* (**R**)
A tree. Available all year. **Uses:** 1) Only a few leaves are eaten by animals because the farmers said it is "not very nutritious." 2) Timber and fuel. 3) Seeds are roasted/fruit.

Cinnamomum tamala (**Buch-Ham**) **Nees & Eberm.; Lauraceae;** *Gobelshi* (**N**); *Shishi* (**N,R**); *Mashita* (**S**)
A small tree which grows for 3-5 years. November to January. Found near Dobrato. **Uses:** 1) Cow and sheep fodder. 2) Spice (cinnamon).

Dalbergia pinnata (**Lour**) **Prain; Papilionaceae;** *Damar*
A shrub used for fodder.

Daphne bholua **Buch.-Ham., ex D.Don; Thymelaeaceae;** *Lokta*(**N**)
An evergreen shrub about 2 m high. **Uses:** 1) Rope is made from the fibrous bark to carry loads. 2) The fibrous bark is sold to make traditional Nepalese paper.

Dendrocalamus hamiltonii; **Gramineae;** *Ban, Choyan*
Used for building material and household items.

Dryopteris cochleata; **Aspleniaceae (Dryopteridoideae);** *Nigro* (**N, R**); *Buquluk* (**S**); *Lhakpa Cherman* (**S**)
Fern-like with yellow green leaves which grows about one meter high. No fragrance. Found by rocks in open space. Gongtala and Yangden. March-May (Sherpas), November-February (Rais). **Uses:** 1) Food, cooked as a vegetable by both Rais and Sherpas. 2) Cow fodder.

Dryopteris fillix ms L. Schott; Aspidiaceae; Uniyo (**N**); *Wasem* (**R**); *Tokcha* (**S**)
A dense hair herb, grows 2 m or more. Available from February-March, but during March-May said to be"very good." Found in Yangden, Pakhala (very small). **Uses:** Animal fodder (Sherpas), when matured.

Elephantopus scaber **L.; Compositae;** *Mulapate* (**N, R, S**);
An herb with green leaves which grows up to 50 cm. A yellowish stem near the root. Dore. June-October. **Uses:** Cow fodder; a little sheep fodder.

Elsholtzia blanda; **Labiatae;** *Van Silam*; *Bubuyam* (**R**); *Purmang* (**S**)
Olive-green leaves. Has a sweet smell, almost like mint. Grows about 2 m high. Gongtala, Dobadak. Around May. **Uses:** A religious plant used only by the Sherpas.

***Ficus nemoralis*; Moraceae; *Dudhilo* (N, R); *Homishing* (S)**

A small deciduous tree with green leaves, a red line in the middle and a red stem. Grows 10-12 m high. Considered "nutritious" by the farmers. Found near Gongtala in the *jangal*. January -May. **Uses:** 1) Fodder for cows, sheep, goats and bulls. 2) Firewood

***Ficus roxborghii*; Moraceae; *Nibara* (N); *Waspo* (R); *Abli* (S)**

A broadleaf tree with large crown (leaves). Found at 1,700 meters at the Gong. From November during the rainy season. **Uses:** 1) Animal fodder, considered very "nutritious." 2) Firewood

***Fragaria vesca* L.; Rosaceae; *Aiselu* (N); *Ghis* (R); *Changnyanima* (S)**

Leaves are forest green and the fruit is similar to an orange. Grows 1.5 m high. Found near the Yangden path and at the Gong. Available: April-June (fruit). **Uses:** 1) Fodder: Sheep and goats. 2) Food

***Gentianaceae*; Asteridae (?); *Bazar Guru* (R,S)**

A herb with green leaves and a yellowish-green stem. Contains a iridoid substance (has a bitter taste). At the *lekh*. June-October. **Uses:** 1) Medicine: The root is used for colds and fever by Sherpas. The Rais use it for cuts.

***Girardinia diversifolia* (Link) Friis; Urticaceae; Himalayan Nettle; *Sisnu* (N); *Gaanam* (R); *Yangok* (S) /yangop**

A perennial herb with stinging fuzzy hair leaves. Found in open spaces near Gongtala and Yangden. April-May (delicious), May-June (food), June-July (rope). **Uses:** 1) Leaves used as a vegetable, soup or in a stew (with corn kernels, or small potatoes). It requires a long time to cook as the leaves are tough. Rais eat it as a vegetable. 2) Stems are eaten by pigs and cows. 3) Used as rope when fully mature. 4) *Allo* made into purses and clothes by Rais. 5) Punish naughty children.

***Holboellia latifolia*; *Lardizabalaceae*; *Gufla* (N); *Bye Bye Pou* (N); *Lugimber* (S); *Kikulu* (S); *Taw* (N); *Ludhumsi* (R); *Lukgyanbur* (S); *Luktimber* (S)**

A woody climber which looks like a chayote and is yellowish-green. Beautiful flower. The ripe fruit is red (black seeds). Found near Dhap, Daorali, Dobrato, Dobatak and Gongtala. Available June-November. **Uses:** 1) Livestock fodder. 2) The fruit is rosy pink on the outside, the inside flesh is white and the seeds are black. The ripe fruit pulp is eaten by both Sherpas and Rais.

***Juniperus incurva*; Cupressaceae; *Dupi* (N, R); *Shukpa* (S)**

A plant with yellow-green needles and a brown stem. Smells like pine. Grows about 1 m in height. Found at the *lekh* (Tin Pokhari). June- September. **Uses:** 1) A religious plant, and is burned as incense. 2) Used occasionally as firewood at the *lekh* because it burns quickly and provides warmth. It helps to dry other fuelwood.

Kobresia nepulensis; **Gramine Grass Family;** *Buki* **(R, S)**

Grows 4-5 in high in open space. Found at Dore, Khola Kharka. Available from June-November. **Uses:** Fodder for all the animals. Considered the most "delicious," and "makes the milk come."

Litsea citrata; **Lauraceae;** *Siltimur* **(N);** *Khokchikapaa* **(R);** *Dhum* **(S)**

Has green leaves and green berries. The fruit turns black when ripe. Grows up to 3 m in height. Nice fragrance. Found in *jangal* close to Gongtala. July-September. **Uses:** 1) Food (young fruit mixed with chili by Rais) or is eaten as a fruit. 2) Medicine for severe diarrhea (dried fruit).

Musa paradisiaca **L.;** *Limchok* **(N);** *Limchasi* **(R);** *Poyang* **(S);** two types

A huge tree. Has a sweet banana. Found near Deorali and Karani. Available during July-September. **Uses:** 1) The fruit is eaten by both Rai and Sherpa adults and children. 2) Used as medicine for cows.

Mussaenda frondosa **L.;** **Rubiaceae;** *Asare* **(N, R);** *Napiling* **(S)**

A shrub with green leaves. Grows between 2-7 m high. Found in open space and *jangal.* Berries are burgundy red when ripe. Very sweet. June-September (berries are ripe). Gongtala, Yangden, May-June, (flowers). **Uses:** 1) Berries are eaten by Sherpa children and adults. Not eaten by Rais. 2) Bird food. 3) Firewood (Rais).

Myrica esculenta **Buch-Ham.ex D. Don;** **Myricaceae;** *Kaphal* **(N, R);** *Dulung* **(S)**

An evergreen tree which grows 2 m high or greater. Yellowish-green leaves with edible small red berries. Between Yangden and Gongtala. Available: fruit (March-May). **Uses:** 1) Berries are eaten by both Rais and Sherpas. 2) Bird food 3) Sheep and goat fodder.

Perilla frutescens **L. Britton;** **Labiatae;** *Silam* **(N);** *Chomba* **(S)**

A tall hairy herb which grows in swidden fields, alongside millet. **Uses:** 1) A condiment, side dish, snack and spice. 2) Pounded or roasted seeds and seed oil used as food.

Persea odoratissima **(Nees.);** *odoraeissima;* **Kosterm; Lauraceae;** *Kaulo* **(N);** *Nichisi* **(R);** *Paiyuma* **(S)**

An evergreen tree which grows up to 5 m in height. During the month of August-September, the flowers are black and are eaten by Rais. Around Dhap, it is available. **Uses:** 1) Eaten by Rais and Sherpas. 2) Cow fodder (December-February). 3) Bear food.

Phyllanthus Emblica **L.;** **Euphorbiaceae;** *Koroshi* **(R, N);** *Chorok* **(S);** *Namcharok* **(S);** **Cherok (S);** **Gundung (S)-Solu Khumbu**

A 2 m tree with yellowish-green feathery leaves, has yellowish-white flowers, and bluish-green sour fruit. No fragrance. Found around Gongtala, Dobatak and

Yangden. July-October (Sherpas) available. December-January (Rais). Available: March/April (medicine). **Uses:** 1) Black fruit eaten by children, September-October. 2) Human food, *khole* (flower and water), make relish with radish. 3) Fodder (leaves): cows and goats. 4) Animal medicine use by the Rais. 5) Diarrhea and stomach ailment medicine used by the Rais (dried berries). Flower is squeezed with cloth for 2-3 hours. 6) Bird food.

Phytolacca acinosa Roxb.; Phytolaccaceae (Poke weed); *Jaringo sāg* (N); *Dumdum sāg* (R); *Danur* (S)
A semi-domesticated perennial leafy herb. Planted, but many years ago with seeds from the *jaṅgal*. April-November available. Found in Gongtala and Dobatak. **Uses:** food, eaten as a vegetable by both Rais and Sherpas.

Picrorhiza scrophulariiflora Pennell.; Scrophulariaceae; *kutki* (N, R); *Hungling* (S)
A small herbacious plant which has purplish-black flowers, green leaves, and a brown root. Found at Tin Pokhari, Dore (*lekh*), 3,000-4,000 m. June-November. **Uses:** Medicine (root): cold, stomach ailments (Rais), headache, cough and fever. It is made into a tea and drunk (added to boiled water). Bitter taste.

Potentilla peduncularis; Rosaceae; *Kukura Pankhi* (N, R, S)
Flowers have a golden color. Grows 4-5 in high in open space. Dore, Khola Kharka. May-November. **Uses:** Cow and a little sheep fodder reported by Sherpas.

Quercus lamellosa Sm.; Fagaceae; *Gogane* (N)
A large evergreen tree. **Uses:** 1) Timber and firewood. 2) Leaves: animal fodder.

Rheum australe D.Don; Polygonaceae; *Kokim* (N, R); *Churcha* (S)
The flowers and the stems are burgundy red. Grows up to two feet. *Lekh*. June-September. **Uses:** 1) Medicine: a) The root is used for fever, and b) Placed on cuts by both Rais and Sherpas. 2) Drunk by the Rais as a tea for body ailments and fever,

Rheum webbianum Royle; Polygonaceae; *Padamchal* (N, R); *Keju* (S)
A perennial herb with pale yellow flowers and leaves, a brown root. Grows about 2 m high. Found around 3,000 m at Dore and at the *lekh* (4,000 m), just below the snowline. June-October. **Uses:** 1) Drunk as a tea only by Sherpas. 2) Used as medicine for fever (a very bitter taste), body aches and cuts by Sherpas only. Dried, ground and made into a paste. 3) Inner stalks are eaten and are slightly sweet. 4) Cow medicine (cooked and given). 5) Traded to Dharan and Hile and exported to India where it is processed into medicine for the international market. Sold for one rupee as medicine at the road bazaar.

Rhododendron anthropogon **D.Don; Ericaceae;** *Sunpati* **(N),** *Dhup* **(R);** *Masur* **(S);** *Po* **(S)**

A small bush with thick green leaves nearly 1 m high. Found in open space and rocky areas. High altitude, *lekh*, approximately 5,500 m. White flowers (August-September). Sweet fragrance. August- September (Surje). **Uses:** 1) Religious plant. Leaves and twigs are burned as incense (Rais and Sherpas). 2) Tea. 3) Sherpas barter with Yaphu Rai: 1 *doko* sunpati for 8 *pāthi* maize (32 kg).

Rhododendron setosum **D. Don; Ericaceae;** *Sunkhera* **(R);** *Kisur* **(S)**

A bush with green leaves and a brown stem. Grows 12-18 in high in open space. Similar to *sunpati*, smells when thrown in fire. Dore, Lemini. May-November. Winter months not available. **Uses:** Incense (Sherpas).

Rubus niveus **Thunb.; Rosaceae; Red raspberry;** *Ghis* **(R);** *Keshhokpa* **(S)**

Brown seed (red berry inside). Grows 2 m high. Available all year in Gongtala and Yangden. **Uses:** 1) Human food. 2) Bird food. 3) Goat and cow fodder.

Saussurea gossypiphora **D.Don (Himalaya); Compositae;** *Tara Phul* **(N);** *Rudopo* **(R);** *Khapal* **(S)**

A dwarf perennial herb with a cotton-like flower head, green leaves and a magenta stem. Grows to 12 in high. *Lekh* (between 3,800-5,200 m). Flowers July-October. **Uses:** Medicine used as plant paste for cuts by Rais and Sherpas.

Strobilanthes goldfussia; **Acanthaceae;** *Kursani* **(N);** *Chanyal* **(S)**

A shrub with spiny olive green leaves grows one metre high. March-November. Around Tokpu. **Uses:** Cow fodder.

Swertia chirata **Roxb.; Gentianaceae;** *Chiraito* **(N);** *Khipli* **(R);** *Tikta* **(S)**

An herbaceous annual plant which grows to 2 m. Flowers are light green in color and tinged with purple. Grows in swidden fallow fields (1,200-3,000 meters). July-October. **Uses:** 1) Medicine for colds, cough, fever and stomach ailments, used by Rais and Sherpas. 2) The entire plant is collected and sold by Rais and Sherpas. During October-November, traded to India where it is produced for the international market.

Swertia multicaulis; **Gentianaceae;** *Sergundrum* **(S)**

A herbaceous plant with green leaves, a yellowish brown root and white flowers. Grows 8-10 in high. June-September. *Lekh*. **Uses:** Medicine(root) for cuts. It is pounded and then rubbed into the cut.

Symplocos sp.; **Symplocaceae;** *Karane* **(N, R);** *Talashing* **(S)**

Grows very tall, similar to a bean vine, yellow-green leaves. Magenta purple on the outside and inside, the white part (except for the seeds). Near to Gongtala. September-November. **Uses:** 1) Bear food. 2) Fruit is eaten by the Sherpas.

Thysanolaena maxima (Roxb.) Kuntze; Graminaea; Amriso; Amlise

A tall grass. **Uses:** 1) The flowers are used as broom. 2) The leaves used for fodder.

Trewia nudiflora; **Euphorbiaceae;** *Gutel* **(N);** *Gural* **(R);** *Senyuk* **(S)**

Found in Yangden. June-August. **Uses:** 1) Fodder (cow, goat, sheep), leaves 2) Young plant used (vegetable dish, added to *shakpa* (Sherpa stew and *sisnu* soup) and eaten only by Sherpas. 3) Building material (house)- when fully mature and tall. 4) Edible fruit

Umbelliferae; **Apiaceae;** *Chayamphul* **(N);** *Tolenta* **(R);** *Komak* **(S)**

Grass, 4-6 inches high with olive-green flower. The root is brown on the outside and when it is peeled, yellow inside. Found in open space. Khola Kharka. June-September. **Uses:** 1) Medicine- the root is used to treat fever. 2) Pounded and eaten with chili seed. Smell is similar to potato. Delicious taste. 3) Insect repellant: the stem is placed inside the woolen clothes. 4) The Flower is eaten as a vegetable, but first it is dried (eaten by both Rais and Sherpas). 5) Ornamental- necklace made by the Rais.

Vaccinium sp.;; **Ericaceae;** *Salesi* **(R);** *Olo malo* **(S)**

Grows together with *Kalinnakchun*(S), which has large leaves and burgundy red berries. It grows on a tree called *pal kalinokchum*. Available during the months of June-September. March-May (the root is available). Dhap and Dobrato available. **Uses:** 1) Medicine: for fever, only used by Sherpas (fruit). 2) Food: Rais eat the fruit.

Zanthoxylum armatum **DC; Rutaceae;** *Timur* **(N);** *Yerma* **(S)**

A small tree about 3 m high. Flowers: April-May. Fruits: July-November. Found in Gongtala. **Uses:** 1) Used as a spice or condiment. 2) Medicinal use, fruit (stomach ailments).

Unidentified Plants:

Aktingok **(S)**

A white fruit (some are red and others are white). Found at Karani and Chonggre during March-May.

Arupate **(N),** *Hangupa* **(S)**

A large tree with black fruit. Found in Dobatak and the Gong. September-October. **Uses:** 1) Firewood 2) Fruit

Bhurmang sāg **(R);** *Bhurmang cherman* **(S)**

Grows on trees or from the ground. Found near the Gong and Yangden (1,500 meters). Available December-February. **Uses:** 1) Cow, sheep and goat fodder. 2)

Jangal: eaten by wild animals. 3) Human food- sherpa stew, vegetable dish and as soup.

Duti Cheru (N); Ribho (R); Womi Chichi (S)

Grows 4-5 in high, later about 2 m. *Lekh* July-December. December-May, a little is available. **Uses:** Cow fodder, "makes the milk come."

Indrini (N); Roktokhom (R); Ajenge (S), Akagingin (S),

A semi-domesticated tree plant, like chayote. Planted once and grows in the *jangal*. Has a brown seed, inside is white and an orange fruit. Available in the Rai village September-November. **Uses:** Food: Eaten as a seed only by Sherpas. Pounded and cooked as a soup, tastes like egg soup, a red color.

Jiblolim (R, N); Kirkim (S)

Available all year. During the winter a lot is eaten. Found below Gongtala. **Uses:** Fodder: goats and sheep.

Jurpala (R); Qurkim (S)

A small plant which grows on rocks and tree trunks, only the leaves and red stem, a little root. Red plant found near Dhap in *jangal*. May-October. **Two types:** 1) Red in color (found at high altitude). 2) White- found close to Gongtala, found on the ground, tree or stone. **Uses:** 1) Pinkish-red stem is eaten, sour, Rais eat it occasionally. 2) Medicine for stomach ailments (Sherpas only).

Kanchirna (N); Song khem (R); Kala Chherpa (S)

Grows very tall on trees and stones. From November-May. Around May. **Uses:** Pig's fodder, a little eaten by cows.

Kejo (R, S)

Brown root found under the ground at the *lekh*. August-September. **Uses:** Medicine: for stomach ailments and wound healing. It is mixed with water and then eaten. Has a very bitter taste. It is not sold.

Lara (N); Ludumshi (N, R); Tau (S)

It climbs another tree and similar to chayote. It's flower looks like a banana. Found near Yangden and Dobatak. Ripens October-December. **Uses:** 1) Fodder for all livestock. 2) Rope, used to tie cows. 3) Used to make houses. 4) Human food

Nakshe (N); Sogo khorsaii (R); Nakima (S)

Grows about 1 m high. Found near Dhap and Dobrato. Available from January to May. **Uses:** 1) The flower is eaten. 2) Made into chili by both Rais and Sherpas. 3) Animal fodder.

Pamugoen (S)

A purple fruit.

Phararausi (R); Kalinakchun (S)

Fruit, burgundy-red when ripe. Found together with *olo malo*. June-September.
Uses: 1) Medicine for fever Sherpas). 2) Fruit eaten by both Rais and Sherpas.

Pipiho (N, R); Jangma (S)

Found below Gongtala. Available all year round. **Uses:** Eaten by all livestock.

Salesi (R); Changgudum (S)

Berries are burgundy-red when ripe. Stem is lime green. Like chayote vine.
Deorali, Dhap. August-September. **Uses:** Food for both Rais and Sherpas.

Samshipo (N, R); Photoshing (S)

Found near Gongtala, Dobatak, and Dobrato. Available: December-March.
Uses: Animal fodder

Wabermang (R,N); Kishuma; Kotama; Choshok (S)

Available from November, but not eaten during the rainy season. Around
Yangden. **Uses:** Sheep and cow fodder.

Waldekpa (S)

Green leafy vegetable. About 1.5 m in height. Found at Dhap and Gongtala.
Available during March-May.

Mushrooms:

**Karne cyāu (N), Ambangshamu (S); Ghog sumung (S), Gokshemung (S) Two
types**

Mushroom, light beige underneath, tannish-brown on top and has a brown
stem. Grows on trees in the *jangal*.

Only available during July-September. Found in *jangal* above Saisima. In
Gongtala is available. May-September. **Uses:** 1) Food, fried as vegetable, eaten by
both Rais and Sherpas. 2)Traded by Rais.

Kālo cyāu (N), opelek (S)

Grows on fallen trees. Rais dry it for later consumption. March-May.

Chinduk chamu (S), chibre

Grows on fallen trees. Found in Gongtala. Available during March-May.

Cilme cyāu (N), cilding shamu (S)

Sources of Identification

APROSC. 1991. *Glossary of Some Important Plants and Animal Names In Nepal.* Agricultural Projects Services Centre (APROSC), Kathmandu: Malla Press.

Chevallier, Andrew. 1996. *The Encyclopedia of Medicinal Plants.* New York: DK Publishing Inc.

Mabberley, D.J. 1993. *The Plant-Book. A Portable Dictionary of the Higher Plants.* Cambridge University Press: Oxford.

Majupuria, Trilok C. and D.P. Joshi. 1989. *Religious and Useful Plants of Nepal and India.* Bangkok: Craftsman Press.

Manandhar, N.P. 1989b. *Useful Wild Plants of Nepal.* Nepal Research Centre Publications No. 14. Stuttgart: Franz Steiner Verlag Wiesbaden GMBH.

_____. 1980. *Medicinal Plants of Nepali Himalaya.* Kathmandu: Ratna Pustak Bhandar.

Negi, S.S. 1992. *Minor Forest Products of India.* Delhi: Periodical Experts Book Agency.

NERRA/IIRR. 1992. Regenerative Agriculture Technologies for the Hill Farmers of Nepal. Kathmandu.

Regmi, Dr. Puskkal P. 1992. Personal communication. Central Agronomy Division, Kathmandu, Nepal.

Shakya, Dr. Puspa R. 1992. Personal communication, Makalu-Barun Conservation Project, Kathmandu, Nepal.

Shrestha, Dr. Tirtha B. 1992. Personal communication, International Union for Conservation of Nature, Kathmandu, Nepal.

Standal, Dr. Bluebell R. 1997. Personal communication, Honolulu, Hawaii.

Toba, Sueyoshi. 1975. Plant names in Khaling. A study in ethnobotany and village economy. *KAILASH* 3(2):145-169.

Wild Edible Plants of Nepal. 1982. *Wild Edible Plants of Nepal.* Department of Medicinal Plants, Nepal, Bulletin No. 9. His Majesty's Government, Nepal.